T0350769

Banking Systems Simulation

Wiley Series in Modeling and Simulation

The Wiley Series in Modeling and Simulation provides an interdisciplinary and global approach to the numerous real-world applications of modeling and simulation (M&S) that are vital to business professionals, researchers, policymakers, program managers, and academics alike. Written by recognized international experts in the field, the books present the best practices in the applications of M&S as well as bridge the gap between innovative and scientifically sound approaches to solving real-world problems and the underlying technical language of M&S research. The series successfully expands the way readers view and approach problem solving in addition to the design, implementation, and evaluation of interventions to change behavior. Featuring broad coverage of theory, concepts, and approaches along with clear, intuitive, and insightful illustrations of the applications, the Series contains books within five main topical areas: Public and Population Health; Training and Education; Operations Research, Logistics, Supply Chains, and Transportation; Homeland Security, Emergency Management, and Risk Analysis; and Interoperability, Composability, and Formalism.

Founding Series Editors:

Joshua G. Behr, Old Dominion University
Rafael Diaz, MIT Global Scale

Advisory Editors:

Homeland Security, Emergency Management, and Risk Analysis

Interoperability, Composability, and Formalism

Saikou Y. Diallo, Old Dominion University
Mikel Petty, University of Alabama

Operations Research, Logistics, Supply Chains, and Transportation

Loo Hay Lee, National University of Singapore

Public and Population Health

Peter S. Hovmand, Washington University in St. Louis
Bruce Y. Lee, University of Pittsburgh

Training and Education

Thiago Brito, University of Sao Paolo

Spatial Agent-Based Simulation Modeling in Public Health: Design, Implementation, and Applications for Malaria Epidemiology

By S. M. Niaz Arifin, Gregory R. Madey, Frank H. Collins

The Digital Patient: Advancing Healthcare, Research, and Education

By C. D. Combs (Editor), John A. Sokolowski (Editor), Catherine M. Banks (Editor)

Banking Systems Simulation

Theory, Practice, and Application of
Modeling Shocks, Losses, and Contagion

Stefano Zedda
University of Cagliari
Cagliari, Italy

This edition first published 2017
© 2017 John Wiley & Sons, Inc.

The right of Stefano Zedda to be identified as the author of this work has been asserted in accordance with law.

Registered Office
John Wiley & Sons, Inc., 111 River Street, Hoboken, NJ 07030, USA

Editorial Office
111 River Street, Hoboken, NJ 07030, USA

For details of our global editorial offices, customer services, and more information about Wiley products visit us at www.wiley.com.

Wiley also publishes its books in a variety of electronic formats and by print-on-demand. Some content that appears in standard print versions of this book may not be available in other formats.

Library of Congress Cataloging-in-Publication Data applied for.

Hardback ISBN: 9781119195894

Cover image: © peshkov/Gettyimages
Cover design by Wiley

Set in 11.5/13.5 pt WarnockPro-Regular by Thomson Digital, Noida, India

10 9 8 7 6 5 4 3 2 1

The book "BANKING SYSTEMS SIMULATION" intuitively presents essentials tools for integrating risks in banks and bank systems that are essential for risk professionals and regulators.

Prof. Harald Scheule,
University of Technology Sydney,
Australia

Stefano Zedda succeeds in illustrating a wide array of state of the art techniques to set up effective models of bank management, risks and contagion. Real life applications show how the book's key concepts can be used to overcome data limitations and develop parsimonious, yet accurate representations of how outside shocks and policy changes affect lenders.

Prof. Andrea Resti,
Università Bocconi,
Italy

The book offers a comprehensive analysis of bank and banking system risks by adopting a simulation framework and an integrated approach between the micro and macro dimension of risk management. The analysis provides important insights for academics, regulators and practitioners.

Prof. Francesco Vallascas,
University of Leeds,
UK

Contents

Foreword

Why write (or read) another book about models of banking? It sometimes seems that banking is passé — that the real financial action lies elsewhere. A recent survey (R. Greenwood and D. Scharfstein, "The Growth of Finance," *J. of Econ. Perspectives*, 2013) documents the explosive growth in non-bank financial products and services. Securities markets more than quadrupled in size, from 0.4 to 1.7 percent of GDP between 1980 and 2007, on the eve of the crisis. Surely this is where we should focus — the busy secondary markets for bonds, equities, securitized products, and over-the-counter derivatives.

Yet, reports of the demise of traditional banking have been greatly exaggerated. Credit intermediation, which includes traditional deposit-taking and lending alongside banks' transactional services such as credit-card accounts and ATM activity, have also grown. Starting from a much larger share, Greenwood and Scharfstein (2013) calculate that credit intermediation grew from 2.6 to 3.4 percent of GDP over the 1980–2007 period. When the dust has settled on a quarter century of remarkable growth, banking is still roughly twice the size of securities markets.

One key reason that banking has been able to keep pace with the booming secondary markets is that banking is a key player in them. Banks provide custodial services for investors and asset managers, prime brokerage services for hedge funds, and much of the loan origination at the front end of asset securitization pipelines. Banks also still dominate the wholesale funding markets that manage much of the financial sector's liquidity provision every day.

Perhaps most importantly, banks provide the crucial cornerstone of financial capital upon which much of the edifice of credit expansion is built. The ability to leverage capital to extend liquidity (by expanding lending) during an economic expansion is critical to the functioning of the system. But the possibility of overleveraging and aggregate liquidity shocks are critical dangers. The crisis of 2007–09 demonstrated that these are not idle fantasies. The challenges of defining "adequate" capital and liquidity levels and ensuring that banks meet this standard are the driving forces behind the often highly technical conversations around the new Basel III, supervisory stress testing, and orderly resolution protocols.

In short, the management and regulation of banks and banking remain important, challenging, and timely topics, worthy of our attention.

Why is simulation a useful approach for addressing these topics? Simulation has two key features that make it an appropriate methodology in this context. First, the financial sector in general, and banking in particular, are evolving rapidly. In addition to the growing volumes of activity, there are significant innovations in both institutions and practice. For example, an entrepreneurial wave of fintech innovation is working to disrupt retail payments and traditional lending channels; supervisors are forging ahead with major new efforts in stress testing and data-driven regulation; and post-crisis institutional reforms are forcing the clearing and settlement of many over-the-counter transactions onto central counterparties. The upshot of these changes is that past is not prologue for many important questions. Moreover, the pace of innovation is such that market participants and regulators alike continue to wrestle with yet new counterfactual proposals. When simple historical patterns cease to be reliable, a higher-order model is vital. Second, many of these issues, especially at the system level, involve intricate interactions and nonlinear feedback effects that pose insurmountable tractability challenges for more traditional theoretical models.

Simulations can be implemented poorly, of course, but when done well, they have the potential to provide us guidance on this turbulent voyage. Books like the one you are reading help move us toward this better practice.

Mark D. Flood
Washington, D.C.
February 2017

Introduction

Simulation methods have recently received great attention, and many studies based on this approach have been developed in the last decade for assessing banking systems stability, determinants, and possible consequences.

Different approaches have been developed to address the main problems, but no single paper aimed at analyzing some specific aspect gives a complete picture or an orderly presentation of the topic.

The aim of this book is to present in an orderly manner the main steps, information sources, and methodologies developed for modeling and simulating banking systems stability and its applications.

The recent financial crises have led to the realization of the importance of simulations, as on one hand systemic risks are really important, and on the other hand it is not possible to make all analyses based on actual data, as the available data are limited by case number, early intervention of supervisors, and the framework evolution.

The modeling and simulation approach, which has been particularly described and developed in this book, is based on a theoretical representation of the fundamental mechanisms of risk managing of banks, and it aims at simulating the possible outcome of the banking sector as a consequence of important shocks, or for having some clue about what the consequences can be in case of regulation changes, modifications in the system structure, introduction or modification of the safety net, and other possible interventions or policies.

This book also includes the main references for banking systems risk simulation, including the models used for representing and quantifying the main banking risk sources, the banking network linkages representation and estimation, correlation and contagion mechanisms, simulation models and methods, and the most important applications for evaluating and testing the effects of possible interventions and regulation changes, and contributes to a better understanding of the banking systems risks and stability.

1

Banking Risk

A bank's core business, credit activity, is centered on borrowing and lending, thus mainly dealing with two components: money and risks.

The first component, money, seems to be the simplest to measure, as all balance sheets and income statements report only money values. In fact, the different contracts, timing, and liquidity require much more attention than expected.

Every economic activity implicitly includes risk, as the economic framework always includes uncertainty. But a bank's activity is centered on risk, as its core business is in borrowing money, and lending it, bearing all risks of counterpart: default, maturity transformation, market values variation, liquidity, and so on.

Diverse layers of bank activity are cross-linked, and take part in maintaining the equilibriums in terms of revenue, economic stability, and operational activity. As a consequence, the bank's activity analysis is always complex.

The ways for analyzing credit activity are multiple.

On one hand, the banking activity is a specification of firms management, so it can be analyzed with the same attitude in terms of internal processes, costs and income, business models, personnel management, and so on.

Another possibility is to evaluate the activity results of banks from the outside, by means of regression analyses, so as to find a posteriori a description of their actual activity, results, and business models distribution and evolution.

Banking Systems Simulation: Theory, Practice, and Application of Modeling Shocks, Losses, and Contagion, First Edition. Stefano Zedda.
© 2017 John Wiley & Sons, Ltd. Published 2017 by John Wiley & Sons, Ltd.

Banks play a key role in financing the real economy, thereby sustaining and promoting the economic growth; their activity is often considered to be of national interest, and in some countries it is directly held by public companies.

In fact, the credit support of a firm or sector can substantially change its evolution and growth; choosing which firm to finance or which sector to support can be in some cases more effective than some public policy interventions.

Other fundamental aspects of banks activity are related to the volume of money managed in stock exchange and bonds markets, where the buying and selling activity can significantly affect values. Even when considering issuers of large dimensions, as in the case of sovereign bonds, the bank's attitude to buying or holding bonds to maturity can be of fundamental importance, and often the interest of governments in keeping the availability of banks in this can sometimes affect the government policy toward the banking sector.

This key role in sustaining the economic growth and the fact that banks are typically large firms induces specific attention toward the bank's activity, as the banks' default not only stops the support of economic growth but also can induce huge effects of market instability, lack of confidence in banks and in savings, bank runs, and disruptive effects on the real economy.

Thus, the analysis of a bank's activity, and of its different layers and interconnections, and the supervision and regulation of banks, are of fundamental importance for preserving savers' confidence in banks, the bank's action in channeling savings to firms, thus sustaining economic growth and preserving economic and financial stability.

The credit activity also carries a specific characteristic, as it involves buying and selling money—different maturity, contracts, risks, but always money. There is no actual goods production or transformation. This simplifies some aspects, but also induces a greater interrelationship between the different activity layers; so, as an example, there can be no strict separation, as it happens in industrial or commercial activities, between real goods or services production, and financial activities.

At a first glance, a bank's balance sheet seems to be quite similar to any other firm's balance sheet: The assets side mainly includes customer loans, bonds, interbank credits, and some other assets such as cash, buildings, and so on. As banks do not buy, transform, or sell goods, there is no motivation for quantifying the values of goods at the beginning and end of each year. The liabilities side includes deposits, interbank debts, issued bonds, and capital.

A more significant difference with respect to nonfinancial activities, such as the industrial activity, appears when comparing the assets side with the income statement: For banks, the total revenue is only a fraction of total assets, while industries typically register sales revenues in value closer to the total assets.

As an example, the FCA 2015 group's consolidated balance sheet[1] reports total assets of €105,040 million, equity of 16,255 (15.5% of TA), and net revenues of €110,595 million (105% of TA), while the Deutsche Bank 2015 balance sheet[2] reports total assets of €1,436,029 million, capital of 45,828 (3.2% of TA), and main income values (interest income, current income, commission income, and other operating income) summing up to a total of €31,086 million, around 2% of total assets.

It is evident that when analyzing a bank's activity, our attention is more on the assets volume than on income. It is worth noting that Germany's GDP for 2015 was estimated to be €3,025,900 million,[3] while Deutsche Bank's total assets in 2015 were about 47% of its home country's GDP.

This assets dimension also explains why risks are so significant in banking activity. Referring to the values above, a reduction of 3.5% in value for FCA assets will reduce the equity value from 15.5 to 12.4% of total assets, while in the case of Deutsche Bank the capital will be completely wiped out.

---

1 Source: Fiat Chrysler Automobiles 2015 Annual Report.
2 Source: Annual Financial Statements and Management Report of Deutsche Bank AG 2015.
3 Source: Eurostat.

Another signal of banks' central role is shown by the number of governments' interventions in rescuing banks during financial crises: Interventions on capital are absolutely fundamental when a large bank is likely to fail, and the cost of non-intervention is typically much higher than the cost of capital injection needed for rescuing the bank.

In fact, not only can the effects of uncertainty in assets and liabilities deeply affect income, and thus banking stability, but also dealing with risks is the basis of the banking activity.

So, even if the primary focus with respect to a bank's activity is toward their assets value, the uncertainty of values intrinsically inherent in the lending activity is the key reference for understanding why banking is almost a synonym for risk management.

1.1 Single Bank Risk

The first reference for analyzing a bank's activity is in considering its balance sheet main values.

(JPMorgan Chase & Co./2015 Consolidated Annual Report)

Starting from the assets side (Table 1.1), the most important exposure of banks is for customer financing, by means of loans.

Loans are the traditional banks' core business, which brings a fundamental part of revenues and carries the most significant risks.

In fact, the main activity of banks consists in evaluating whom to lend money, how, and how much to lend. Analyzing a firm's balance sheets, cash flows, and tendencies (hard information), or verifying the firm's reputation, management capabilities, and reference market stability (soft information) are some of the important ways of evaluating the firm's credibility: that is, if there is a strong probability that the firm will meet its obligations and pay back the debts as scheduled.

It is evident that this evaluation cannot be exact. On one hand, it depends on future events that are not possible to forecast exactly, and moreover speculating the reactions of the firm management on these unforeseeable events will be even more difficult. On the other hand, it is not possible to analyze in depth all the firm's aspects and details, and this intrinsically results in

Table 1.1 Bank balance sheet: assets.

Assets	
Cash and due from banks	20,490
Deposits with banks	340,015
Federal funds sold and securities purchased under resale agreements	212,575
Securities borrowed	98,721
Trading assets	343,839
Securities	290,827
Loans	837,299
Allowance for loan losses	13,555
Loans, net of allowance for loan losses	823,744
Accrued interest and accounts receivable	46,605
Premises and equipment	14,362
Goodwill	47,325
Mortgage servicing rights	6,608
Other intangible assets	1,015
Other assets	105,572
Total assets	2,351,698

widening the confidence intervals of the creditworthiness estimation.

As a consequence, it is fundamental for banks to use all possible strategies to reduce the total risk of the lending activity.

The traditional, and still fundamental, strategy is based on diversification. In fact, if the exposures are affected by different risk sources, the total risk is lower than the sum of individual risks. In practice, this means that it is unlikely that all exposures will default at the same time; instead, a good diversification ensures that the fraction of defaults tends to remain near the expected value. In this way, it is possible to maintain the bank's financial stability covering the expected value of defaults by means of interest spreads, and store a capital buffer for possibly absorb losses when its value is higher than expected.

It is worth noting that the bank risk is due to the uncertainty of loss value, and not due to its intrinsic value. For clearer evidence, we can consider the example of two banks of the same size, $100 million—the first exposed to firms with higher default probability, say 10%, and strong diversification (or other risk covering), so the total loss variance is of 2%; the second exposed to less risky firms, with a default probability of 5%, but no diversification (or other risk covering), so a higher variance in total losses, say of 5%. In the first case, there are expected losses of 10 and an uncertainty of 2, while in the second we have expected losses of 5, but an uncertainty of 5. The second is much more exposed to risk, even if the first bank's exposures are for riskier firms, and the expected value of losses is higher, as the second case is more subject to uncertainty.

The second important value in our simplified representation of a bank's balance sheet is for bonds, either held to maturity or for trading.

In traditional bank activity, bonds were one way of lowering the average risk, as typically bonds are issued by large firms or by governments (sovereign bonds), so the risk of counterpart default is typically lower, and as bonds are traded on financial markets, they also have a liquidity reserve role, fundamental for covering unexpected cash needs. Evidently, the lower the risk, the lower the expected income on these investments.

More recently, and in particular for large banks, the trading activity has had an important evolution, visible in the balance sheet as a movement from bonds "held to maturity," to bonds "held for trading." Evidently, this activity is really different from the traditional banking activity, as it is aimed not at financing an investment, but at having an income in buying and selling bonds (or shares, or derivatives) so as to profit from a price differential, thus much more similar to the commercial activity. This kind of operation is mainly exposed to market risk (in addition to the counterpart risk, always present).

Another fundamental value in our representation refers to interbank loans. In fact, banks often lend money to other banks, here also for liquidity management, for investing some momentary money excess, or for covering some momentary cash need. But it can also be due to a specific business model, for which some banks attract deposits, only using part of this savings volume for direct lending, while some prefer instead to invest in the interbank wholesale market, thus concentrating their activity on lending. This role distinction in some countries is between different bank categories, while in other cases it is just a role distinction within a banking group. Banking groups also tend to have a centralized treasury/liquidity management, so that the interbank lending within the group is typically much higher than the lending outside the group.

With reference to liabilities (Table 1.2), the main funding source is in deposits, typically available at sight or on short-term contracts, which provide the bank the funds, but also introduce a mismatching between funding and lending, typically lent with higher maturities.

The stockholders' equity includes common and preferred stock, and retained earnings and other capital reserves, which represent two main sources of bank capital: the issuing of new shares and the retaining of (part of) the earnings produced by the banking activity. The equity is the main shock absorber for banks, and its value is the first reference for limiting the risk of bank default.

The other side of a bank's activity includes reporting the income statement (Table 1.3). It is typically presented starting from the interest income, interest costs, and deriving the interest differential, the net interest income. This value is then corrected for considering the provisions for credit losses. The second layer includes commissions and trading activity for having the noninterest income.

The third layer is mainly devoted to operational costs, but also includes the other values that sum up to the total noninterest expenses. The taxes are then computed for obtaining the net income.

The final result of all the activity is kept by the net income (or loss).

As is well known from the accounting standards, the net income can be distributed among the shareholders, or stored

Table 1.2 Bank balance sheet: liabilities and equity side.

Liabilities	
Deposits	1,279,715
Federal funds purchased and securities loaned or sold under repurchase agreements	152,678
Commercial paper	15,562
Other borrowed funds	21,105
Trading liabilities	126,897
Accounts payable and other liabilities	177,638
Beneficial interests issued by consolidated variable interest entities	41,879
Long-term debt	288,651
Total liabilities	2,104,125
Stockholders' equity	
Preferred stock	26,068
Common stock	4,105
Additional paid-in capital	92,500
Retained earnings	146,420
Accumulated other comprehensive income	192
Shares held in restricted stock units (RSU) trust, at cost	(21)
Treasury stock, at cost	(21,691)
Total stockholders' equity	247,573
Total liabilities and stockholders' equity	2,351,698

for raising the capital value; however, if the bank registers a loss, it must be accounted as a reduction of the capital value.

An important detail with reference to the bank's activity reporting is that some categories of financial investments (in particular, the change in value of "available for sale" investments) are directly imputed on equity, so they do not affect the total and net income, but impact the final equity value. Thus, when evaluating the bank's activity result, it is necessary to reconcile the two, as is done by some commercial databases like

Table 1.3 Bank income statement.

+	Interest and similar income	
−	Interest expense	
=		Net interest income
−	Provision for credit losses	
=		Net interest income after provision for credit losses
+	Commissions and fee income	
+/−	Net gains (losses) on financial assets/liabilities	
+/−	Other income (loss)	
=		Total noninterest income
−	Compensation and benefits	
−	General and administrative expenses	
−	Other noninterest expenses	
=		Total noninterest expenses
=		Income (loss) before income taxes
−		Income tax expense
=		Net income (loss)

Bankscope that specifically takes it into account in its "Fitch Comprehensive Income" value (see Andrew Fight, *Understanding International Bank Risk*, Wiley Finance, 2004).

The banking activity includes several fundamental mechanisms, which are briefly presented here.

The first one is the money channeling from the actors and sectors with more money than needed, mainly depositors but also bondholders or other banks, to the sectors investing in economic activities and producing an income sufficient to pay both the debt and interest.

The evaluation of the firm's ability to pay back debts, so as to have sufficient income and to afford the evolution of the economic framework, is the most important and specific activity of the bank.

This funding transfer needs some specific attention, as the depositors typically have the right to withdraw all of their own deposits without notice, even if they normally need only a fraction of their current account values. So, the bank has to properly quantify which fraction of deposits must be kept available as cash, and which part can be invested in loans or other interest-bearing activities. This quantification is fundamental, since if the cash requests from depositors are higher than the available cash, the bank has to sell some of its activities to obtain the money. But as the ability to evaluate the counterpart creditworthiness is complex, estimating the value of a loan is not simple, and so selling loans contracts in a short time often results in fire selling, thus losing part of the expected interest income. Thus, on one hand, banks continuously monitor the total amount or deposits and cash needs, and, on the other hand, part of their investments is in "liquid" assets, typically in highly traded bonds that can be easily sold at reasonably stable prices.

This is due to another specific aspect of a bank's activity: the maturity transformation. In fact, the liabilities, and mainly deposits, have a short maturity (days), while investments, and loans in particular, are often characterized by a longer maturity (years). This side of the bank's activity is fundamental for the real economy financing, as firms not only need money for financing their investments, but also need it for all the planned investment time. Thus, the bank has to deal with possible deposit volume changes over time, so it is fundamental to attentively monitor the turnover of investments and the related cash needs, so as to maintain the equilibrium in their assets–liability management.

So, on the basis of the actual ongoing deposits volume and stability, banks have to continuously adapt both the volume of high income and high maturity investments (loans) and the low income but highly liquid investments (as sovereign bonds).

Liquidity management can be crucial for banks, as the lack of liquidity is one of the causes of loss of confidence in banks by depositors, which can cause bank runs.

A bank run, even when not justified by actual difficulties in the bank, forces the bank to sell first the liquid assets, and then, if it is insufficient to cover the cash request, to fire sell the high income investments, resulting in important losses, and possibly causing the bank to default. If caution is not exercised, even the false suspicion that a bank is likely to fail can cause a bank run, which can cause the bank eventually to fail!

For this reason, in almost all countries a deposit guarantee scheme is implemented so that depositors may know that their deposits are covered by a guarantee, and any rumors of a possible bank failing would cause less worry and nervous reactions.

Other possible sources of risk are related to the difference between the contracts held on two sides of the balance sheet. For example, as the liabilities side is typically oriented to shorter maturities, it is more exposed to the variability of the interest rate, while the assets side quite often is oriented to fixed interest rates. This mismatching between assets and liabilities brings an interest rate risk, which can be a source of income, when the average interest rates differentials are higher for the bank, but also a possibility of suffering from significant losses in case the floating rate goes above the fixed one.

Similar problems are related to currencies and also to other risk sources.

So, a bank's soundness must be evaluated on different sides, as in the FDIC approach that includes the consideration of *c*apitalization, *a*ssets quality, *m*anagement capabilities, *e*arnings, *l*iquidity, and *s*ensitivity, commonly known as CAMELS.

Finally, banks have to hold equilibriums on different layers, but banks' problems quite often originate from income difficulties or from losses caused by risk exposures, which only become evident later as liquidity problems. In fact, a low income induces banks to take higher risks, such as raising the share of high income operations, with higher rigidity, thus with more exposure to liquidity shortage, while on the liabilities side the low income can induce a reduction in confidence by possible lenders and difficulties in funding.

The risk management activity is evidently of fundamental importance for banks, and needs a detailed and continuous attention for maintaining subtle equilibriums on each of the different risk factors, each one affecting the others. One complete and detailed description of the techniques adopted with this aim is in Resti and Sironi (*Risk Management and Shareholders' Value in Banking*, Wiley Finance, 2007).

If, instead, we analyze a bank's results as the outcome of the whole activity without detailed information on each contract, we can analyze its distribution and evaluate the actual capability of the bank to manage and control the risk equilibriums and, subsequently, to assess the bank's default risk.

If we look at a time series of bank results, we will have something like that shown in Figure 1.1.

In Figure 1.1a, we present the time series of bank results in terms of profits or losses. In Figure 1.1b, we report the frequencies for each percentage, so 0% occurs in year 4, 6, and 8, with a frequency of 3, while −1% only happens once in year 5.

The graph in Figure 1.1a, rotated, gives the standard frequency representation of the profit/loss distribution reported in Figure 1.2.

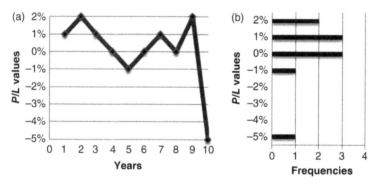

Figure 1.1 Profit/loss (P/L) values and frequencies.

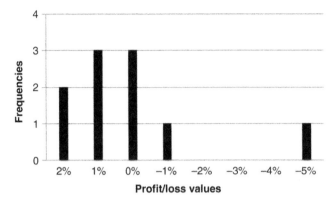

Figure 1.2 Profit/loss frequencies.

This distribution is due to the risks affecting the bank's activity, which on the right side is bell-shaped with a maximum possible loss given by the total value of exposures, while on the left side, as the minimum value of losses is zero, the shape is different. Depending on the exposures riskiness, the left side can decrease from the origin, in case of low riskiness, or it can first increase and then decrease, in case of higher riskiness, similar to what happens in a Poisson distribution depending on the expected frequency of the event.

The leftmost part of the probability distribution, up to the expected value, is covered on provisions, and included in the income equilibrium as a cost, similar to what happens for interest expenses. So, when evaluating the actual losses of the banking activity, the reference is to the expected value, positive numbers represent the value of losses exceeding the expected value, and negative numbers are for the cases where the value of losses is lower than expected.

Following the nomenclature of Demirguc-Kunt (1989), a bank is defined to be economically insolvent when the present value of its assets, net of implicit and explicit external guarantees, falls below the present value of claims from the banks' creditors.

Banks are considered to be in default when losses exceeded capital:

$$L_i > K_i \tag{1.1}$$

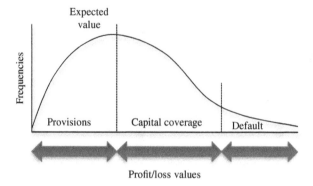

Figure 1.3 Profit/loss probability distribution.

The rightmost part of Figure 1.3 refers to the case when losses are higher than the capital coverage; so, if there is no recapitalization, the bank fails.

This representation explains why the capital coverage has acquired a central role in banking regulation. The Basel Committee on Banking Supervision, since the "Basle Capital Accord" of July 1988, formally "International Convergence of Capital Measurement and Capital Standards," has centered attention on the evaluation of banks' default risk on two fundamental aspects: the assets riskiness evaluation, fundamental for approximating the losses distribution, and the capital coverage, as a main barrier for containing excess losses and limiting the default risk.

1.2 The Basel Committee on Banking Supervision Approach to Regulation

The Basel Committee on Banking Supervision has its origins in the financial crisis that followed the breakdown of the Bretton Woods Accords. After the collapse of the Bretton Woods system of managed exchange rates in 1973, and the following financial market turmoil, many banks suffered from large foreign currency losses. In response, the central bank governors of the G10 countries decided to set up a forum for regular cooperation between its member countries on banking supervisory matters. Consequently, at the end of 1974, they established a Committee on Banking Regulations and Supervisory Practices, located at the Bank for International Settlements in Basel, and later renamed the Basel Committee on Banking Supervision (BCBS).

Since then, the Basel Committee has been the primary global standard-setter for the prudential regulation of banks. Its mandate is to strengthen the regulation, supervision, and practices of banks worldwide with the purpose of enhancing financial stability.

The Committee develops and proposes new supervisory standards and guidelines, and recommends sound practices, seeking endorsement from the Group of Governors and Heads of Supervision (GHOS). Its proposals have no legal power but are provided in the expectation that individual national authorities will implement them.

One fundamental aim of the Committee since its start was to reduce the differences in international supervisory coverage and standards. A first step in this direction was the paper known as the "Concordat" issued in 1975 that set out principles for supervisory standards on banks' foreign branches, subsidiaries, and joint ventures.

In the early 1980s, capital adequacy became one of the main focuses of the Committee's activities. After the convergence on a weighted approach to the measurement of risk, the "Basle Capital Accord" of 1988, also called Basel I framework, established minimum levels of capital for internationally active banks, in order to strengthen the soundness and stability of the international banking system.

In order to assess the capital adequacy of banks, the Committee introduced a weighted risk ratio, in which capital is related to different categories of on- and off-balance sheet exposures, each one weighted according to broad categories of relative riskiness. Considering that for most banks the major risk source is credit risk, even if many other kinds of risk are present—such as interest rate risk, exchange rate risk, concentration risk—the framework was mainly focused on credit risk.

1.2.1 The Basel I Framework

Within Basel I, all banks' assets are classified into five categories, each one with a defined weighting, from 0 to 100%.

In brief:

0% weighting includes cash, and claims on central governments and central banks in national currency or equivalent;

0, 10, 20, or 50% weighting (at national discretion) is for claims on domestic public sector entities, excluding central government, and loans guaranteed by or collateralized by securities issued by such entities;

20% weighting mainly includes claims on multilateral development banks, on banks incorporated in the OECD, or on securities firms subject to comparable supervisory and regulatory arrangements, or on non-OECD banks with a residual maturity of up to 1 year;

50% weighting is for loans fully secured by mortgage on residential property;

100% weighting is for claims on the private sector, fixed assets participations, and all other assets.

In the same approach, all off-balance sheet activities are classified into broad categories as follows:

100% credit risk conversion factor for the activities that substitute for loans (e.g., general guarantees of indebtedness, bank acceptance guarantees, and standby letters of credit serving as financial guarantees for loans and securities).

50% credit risk conversion factor for certain transaction-related contingencies (e.g., performance bonds, bid bonds, warranties, and standby letters of credit related to particular transactions), and for commitments with an original maturity exceeding one year.

20% credit risk conversion factor for short-term, self-liquidating trade-related contingent liabilities arising from the movement of goods (e.g., documentary credits collateralized by the underlying shipments).

0% weight for shorter-term commitments or commitments that can be unconditionally cancelled at any time.

Capital consists of different accounting components. Within the Basel framework, these are grouped into two categories:

Tier 1, the first and more reliable component in terms of loss-absorbing capacity, includes paid-up share capital and disclosed reserves.

Tier 2 includes undisclosed reserves, asset revaluation reserves, general provisions and loan loss reserves, hybrid (debt/equity) capital instruments, and subordinated debt.

The weighted sum of the assets categories values gives the risk-weighted assets (RWA) value, and the minimum capital requirement for banks is set to 8% of the RWA (of which the core capital element will be at least 4%).

Table 1.4 reports an example of RWA computation based on balance sheet values.

Table 1.4 Computation of risk-weighted assets and minimum capital requirement under Basel I.

	Value	Weighting (%)	RWA
On-balance sheet			
Category 1	12,000	0	0
Category 2	54,500	20	10,900
Category 3	31,000	50	15,500
Category 4	101,500	100	101,500
Total on-balance sheet			127,900
Off-balance sheet			
Category 1	0	0	0
Category 2	0	20	0
Category 3	0	50	0
Category 4	5,000	100	5,000
Total off-balance sheet			5,000
Total risk-weighted assets			132,900
Minimum capital requirement		8	10,632

The Basel I approach had a substantial success, so the Committee continued its job and evolved the analysis that led to an important evolution in particular in the assets riskiness evaluation.

In January 1996, the Committee issued the so-called *Market Risk Amendment to the Capital Accord,* designed to incorporate within the Accord a capital requirement for the market risks arising from banks' exposures to foreign exchange, traded debt securities, equities, commodities, and options.

1.2.2 The Basel II Framework

The *Revised Capital Framework* released in June 2004, generally known as "Basel II," revised the whole framework, so as to have a more complete and more accurate mapping of risks and capital coverage. The new framework comprised the following three pillars:

I) minimum capital requirements, which sought to develop and expand the standardized rules set out in the 1988 Accord;

II) supervisory review of an institution's capital adequacy and internal assessment process; and

III) effective use of disclosure as a lever to strengthen market discipline and encourage sound banking practices.

Basel II also widened the risk categories considered for minimum capital requirements (MCR) and, beside credit risk and market risk, has added the operational risk component.

Counterparty risk refers to the possibility that the counterparty will not pay back its debts. This refers not only to loans but also to bonds (including sovereign bonds), interbank lending, and traded assets.

Market risk refers to the risk of selling an asset at a lower price than it was bought for. It only refers to traded assets, is limited to the price variations, and does not include the risk that the issuer of the traded assets defaults, the latter being incorporated in the counterparty risk.

Table 1.5 Computation of capital requirement and solvency ratio under Basel II.

Credit risk	257.1
Market risk	13.1
Operational risk	22.9
Total risk-weighted assets	293.1
Minimum capital requirement	23.5
Actual capital	38.6
Solvency ratio	13.2%

Operational risk refers to all technical problems that might arise, such as robberies, frauds, technical failures, errors, and so on.

In a typical medium-sized commercial bank, counterparty risk accounts for about 80–90% of the overall risk, 5–10% is for market risk, and 5–10% is for operational risk.

Another fundamental innovation of the Basel II framework is that the credit risk weighting is based no more on the assets or off-balance sheet categories, but on the single exposure risk quantification, in terms of probability of default (PD) and loss given default (LGD).

Thus, the minimum capital requirement and solvency ratio are provided in Table 1.5.

Evolving the Basel I 8% fixed rate of capital coverage, within Basel II, the minimum capital requirement for credit risk can be computed on the basis of several different approaches.

Under the standardized approach to credit, risk will be to measure credit risk in a standardized manner, supported by external credit assessments.

Under the internal ratings-based (IRB) approach, banks are allowed to use their internal rating systems for credit risk, with two different options.

Under the foundation approach, banks provide only their own estimates of PD, the other parameters being based on supervisory estimates, while the advanced approach allows banks to internally estimate PD, LGD, and maturity, on the basis of several regulatory constraints and subject to the supervisor explicit approval. For both the foundation and advanced approaches, banks must always use the risk-weight functions provided by the Basel II framework for deriving capital requirements.

The function for computing the unitary capital requirement for each exposure is given by

$$
C = \left[\text{LGD} \times N \left[\sqrt{\frac{1}{1-R}} N^{-1}(\text{PD}) + \sqrt{\frac{R}{1-R}} N^{-1}(0.999) \right] \right.
$$
$$
\left. - \text{PD} \times \text{LGD} \right] \times \frac{1 + (M - 2.5) \times B}{1 - 1.5 \times B} \times 1.06
$$

(1.2)

where the maturity adjustment B is given by

$$
B = [0.11852 - 0.05478 \ln (\text{PD})]^2
$$

(1.3)

and correlation R is proxied by

$$
R = 0.12 \frac{1 - e^{-50 \times \text{PD}}}{1 - e^{-50}} + 0.24 \left[1 - \frac{1 - e^{-50 \times \text{PD}}}{1 - e^{-50}} \right] - 0.04 \left[\frac{S - 5}{45} \right]
$$

(1.4)

depending even on the loan LGD, maturity (M), and firm size (S).

Under the foundation approach, the reference value for LGD is 45% for claims on corporates, sovereigns, and banks.

1.2.3 Credit Counterparty Risk

As acknowledged by the BCBS since its first version of the Capital Accord of 1988, for most banks the major risk source is the counterparty risk, hence the risk that the counterparty will not make the required payments.

As already mentioned, the evaluation of the counterparty capability (and willingness) to pay debts is the core activity of a bank. In fact, a large share of banks' assets is on customer credit, which requires an attentive evaluation of to whom, how, and how much money to lend.

The first question, whom to lend money to, refers to the ability of the customer business activity to produce sufficient income to remunerate the investment, and refund the loan and interests in the agreed time. It is thus evident that the answer cannot be separated from the other two, how and how much to lend, as the amount and conditions are part of the evaluation of the whole investment or business planning.

For the creditworthiness evaluation, a first fundamental step is in knowing and understanding the firm: on the one hand the management, and its capability to plan the business, to have adequate feedbacks for identifying the business problems and opportunities, and adapt the activity to the changes in the market and technical conditions; on the other hand it is funda-mental to have adequate information on the firm ability to produce income, cash flows, financial equilibrium, and business plans and projects.

The amount of funding to be agreed by the bank is another fundamental evaluation, to be attentively tailored on the specific firm and investment plan conditions, as it must be sufficient for the firm to actually realize and manage the investment and business activity, but commensurate with the firm's income capabilities.

Finally, the contract conditions enable the firm to be better able to realize the investment or business, and for the bank to have adequate information on the planned investment or business.

When financing a plant building, it is normal to finance it on the basis of the work progress. So, to monitor the actual building process, and agree for a loan payback with an adequately long time for starting the new production and benefit from the resulting income, or when financing a higher trading volume for a commercial activity, more attention will focus on the exposure flexibility and on the use of the current account for the bank to monitor the ongoing business, and so on.

Another point, with reference to the contract conditions, refers to collaterals and guarantees: in case the direct obligor is not able to pay their debt, some goods are sold for it (collaterals, quite often bonds or goods with high saleability) or the guarantee that some person or firm (with higher credit-worthiness) pays for it. Collaterals and guarantees cannot be the basis of the creditworthiness evaluation, but can help provide a better estimation.

These focuses on the contract conditions are important for a good approximation of the actual risk, but the most relevant attention for the subsequent modeling of the bank risk is on the costs, that is, interest rates and spreads.

The basic criterion is a risk-based pricing, so the higher the risk, the higher the price.

In fact, the bank net interest income equilibrium is based on the difference between interest total income and interest costs plus loan losses. This can also be seen as the difference between interest income minus loan losses and interest costs.

In the latest representation it is more evident that the loan losses and interest income are to be considered as the outcome of the lending activity, and for each risk category of customers there must be a reasonable net positive result so as to cover the interest and operating costs of the bank.

Thus, if the expected losses of one risk category are higher than another, the interest rate or spread must be higher than the other so as to compensate the higher losses costs.

In addition, as the higher default risk includes not only higher losses but also, ceteris paribus, a higher variability in the actual losses, a higher flexibility (with lower income) is needed on the other assets of the bank. So, high-risk exposures for balancing the bank income equilibrium have to equilibrate not only the expected losses value but also the higher variability, and the key variable here is the loan pricing.

The source of information for assessing the firm creditworthiness is often classified into two main categories: hard information and soft information. Hard information refers to easily verifiable data such as income statements, balance sheets, and credit ratings. Hard information, for its high objectiveness, is often the basis of the information exchange between the different responsibility levels, so it is more useful and important for large banks.

Soft information is instead often the basis of relationship lending, and includes all the subjective evaluations, such as management capabilities, their honesty, how they react under pressure, the organizational flexibility, and so on. Relationship lending is based on multiple interactions between bank and borrower, on proprietary information coming also from other firms (suppliers, customers, and competitors), and from the current account or other financial information passing through the bank. Soft information and relationship lending is typically more important for local banks and when dealing with small and medium-sized enterprises (SME).

After the evaluation of the creditworthiness of single firms, the bank's attention must be related to the loans portfolio. As already mentioned, diversification is a highly powerful tool for reducing the portfolio riskiness, given the single customer riskiness.

Table 1.6 shows the effect of the correlation on the sum of two sectors results. Even if the results of the two sectors are the same

Table 1.6 Correlation effects.

Correlation	−0.25			0.79		
	Sector 1	Sector 2	Total	Sector 1	Sector 2	Total
	−35.1	−49.6	−84.7	−35.1	−210.9	−246.0
	−40.0	311.1	271.1	−40.0	−271.4	−311.4
	−63.4	493.9	430.5	−63.4	−538.6	−601.9
	−19.4	241.7	222.3	−19.4	−77.2	−96.6
	−35.0	−149.7	−184.7	−35.0	−149.7	−184.7
	−10.7	362.4	351.7	−10.7	241.7	231.0
	86.5	−210.9	−124.4	86.5	362.4	448.9
	−101.9	−271.4	−373.3	−101.9	−49.6	−151.5
	6.9	−538.6	−531.7	6.9	311.1	318.0
	110.6	−77.2	33.4	110.6	493.9	604.5
Mean	−10.1	11.2	1.0	−10.1	11.2	1.0
Variance	4,187	107,580	101,316	4,187	107,580	145,289

in the two cases, just differing for the order, as in the left side the values are less correlated, the sum variance is lower (101 instead of 145), so the uncertainty in the bank results is lower.

The correlation effects are even more evident when considering more sectors, or, in general, more diversified sensitivities to the risk sources. The higher the distinction, the lower the correlation and the higher the stability of bank results.

What is surely important is to limit the concentration of exposures highly influenced by some specific risk source. This problem, often termed concentration risk, can be due to a large number of exposures to one foreign country, or to a sector highly sensitive to some raw material prices (such as oil prices), or to specific currencies. Evidently, the worst case in this sense is to have a large exposure to a single firm or counterpart, as this can possibly threaten the bank stability in case of a single counterpart default.

Anyway, it is not possible to remove all correlations, as firms are exposed to a number of common risk sources. As an example, with reference to a domestic panel of firms, a change in the tax rate will affect all firms. More generally, firms are exposed to the same economic framework, even if each firm acts in a different way, and so is differently exposed to each economic variable. Hence, the firms' results are almost partially correlated with the macroeconomic framework, and part of the bank risk cannot be excluded by means of diversification.

It is worth considering that, as losses are the possible effect of risk, and thus have an intrinsic uncertainty, banks cannot base the pricing on the actual losses, but on its expected value. This value is determined on the basis of the assets riskiness, and the amount is the reference for the "loan losses allowance" imputed in the balance sheet for correcting the loans value, and of the "loan loss provisions" in the income statement for correcting the interest income.

Within the Basel II framework, no specific methodology is suggested for assessing credit risk. Banks should adopt a sound methodology able to correctly address risk assessment policies, and include procedures and controls for identifying problem loans and determining loan loss provisions.

For an in-depth analysis of the credit risk theory and techniques, see Darrell Duffie and Kenneth J. Singleton, *Credit Risk: Pricing, Measurement, and Management*, Princeton University Press.

1.2.4 Market Risk

In addition to the possible default of the borrower, which is classified as a credit risk, banks are exposed to another risk source related to the assets prices variability—the risk of selling items at a value lower than expected. This can happen when selling positions before their maturity.

In fact, banks hold securities and other financial instruments with two possible horizons. The banking book includes securities that the bank intends to hold up to their maturity, classified as "held to maturity" (or even "available for sale"), for maintaining the liquidity and diversification equilibrium in the banking activity. Instead, the trading book consists of positions in financial instruments and commodities held either with trading intent or in order to hedge other elements of the trading book. It refers to a different activity, trading in financial instruments, that in the last decades has increased its importance in bank balance sheets such that in some cases it becomes the main activity.

Different from the banking activity, where contract conditions, in particular costs and maturities, are fixed before the actual contract starts, so there is a reference value at maturity, the trading activity involves buying and selling at market conditions of financial instruments, so that there is a higher uncertainty on the actual profitability of each transaction, which depends on the actual market conditions.

It is worth noting that also the trading book is exposed to the counterparty risk, as the risk that the counterparty will default in intermediate payments (coupons), or that the firm will go bankrupt (shares and securities), is always present. With reference to this, in particular for foreign exchange risk and commodities risk, market risk is connected to counterparty risk, as the market prices typically internalize the default risk.

Market risk is defined as the risk of a negative impact of adverse fluctuations of financial instruments (on- and off-balance sheet positions) arising from movements in market prices. The risks refer to three main sources:

- *Interest rate risk:* The risk of a change in the fair value of a financial instrument or the future cash flows from a financial instrument due to a change in interest rates, or from movement in the credit spreads for indices or issuers; variations in the interest rate directly affect the variable-income securities by means of its coupons, which will be higher if the interest rate rises, while for fixed-income securities the effect is on prices, the interest rate rise causing a reduction in the security price, proportional to its residual duration. Interest rate risk is measured by means of either the duration or maturity of contracts.
- *Foreign exchange risk:* The risk of a change in the fair value of a financial instrument due to a change in exchange rate or gold price. This evidently applies to financial instruments denominated in foreign currencies, but also to financial instruments sensitive to it, similar to equity derivatives on firms highly exposed to specific markets. Gold is treated as a foreign exchange position rather than a commodity because its volatility is more in line with foreign currencies, and banks manage it in a similar way to foreign currencies.
- *Commodities risk:* The risk of a change or volatility in the price of commodities, for example, agricultural products, minerals (including oil), and precious metals, but excluding gold, or commodity market indices. This also applies to financial instruments sensitive to it, such as equity derivatives on firms highly exposed to these commodities, for example, energy industries to oil prices. The price risk in commodities is often more complex and volatile than that associated with currencies and interest rates, and this can make price transparency and the effective hedging of commodities risk more difficult. In fact, commodity markets are typically less liquid than those for interest rates and currencies, so the changes in supply and demand can have a higher impact on prices and volatility.

The most important and often-used valuation methodology for market risk is based on "marking-to-market."

Marking-to-market is the regular (daily) valuation of positions at market prices (or quotes from several independent reputable brokers). As these values are easily observable and objective, regulation often states that banks must mark-to-market as much as possible. Where marking-to-market is not possible, banks may mark-to-model, i.e. calculate the position value indirectly from a market input, by means of an appropriate model.

The central element of the market risk measurement system is the value at risk (VaR). VaR can be defined as the maximum theoretical loss on a portfolio in the event of adverse movements in market parameters, over a given time frame, and for a given confidence level. Standard references (Basel II) for market risk VaR are a confidence level of 99%, and a time frame of 1 day using 1 year of historical data.

No particular type of model is prescribed by the Basel II standards; banks are free to use models based, for example, on variance–covariance matrices, historical simulations, or Monte Carlo simulations.

In this way, the market risks in trading activities can be monitored on a daily basis by quantifying the estimated maximum level of loss in 99 out of 100 cases, after inclusion of a number of risk factors (interest rate, foreign exchange, asset prices, etc.). The intercorrelation of such factors affects the maximum loss amount.

Each bank must meet, on a daily basis, a capital requirement expressed as the higher of (a) its previous day's value-at-risk and (b) the average of the preceding 60 business days' value-at-risk, multiplied by a multiplication factor set by individual supervisory authorities on the basis of their assessment of the quality of the bank's risk management system, with a minimum of 3.

In 2009, the BCBS, considering that the VaR capital charge computed at 99% threshold was not considering significant large daily losses occurring less frequently than two to three times per year, introduced new standards based on a 99.9% confidence interval over a capital horizon of 1 year, coherently with the whole framework, that reflects a 99.9% soundness standard.

1.2.5 Operational Risk

Operational risk is defined, in the Basel II framework, as the risk of loss resulting from inadequate or failed internal processes, people and systems, or from external events. This includes legal risk, but excludes strategic and reputational risk.

Operational risks are usually not part of the standard "risk management" activity, as these risks are not diversifiable and cannot be managed within the bank credit activity. Instead, as it refers to processes and structures, operational risks can be estimated and, in some cases, reduced, but not fully eliminated, and as long as people, systems, and processes remain imperfect, operational risk must be considered.

Three main methods are accepted for calculating operational risk capital charges, in a continuum of increasing sophistication and risk sensitivity:

1) The Basic Indicator Approach simply requires that banks hold capital for operational risk equal to 15% of the average over the previous 3 years of positive annual gross income.
2) In the Standardized Approach, the capital charge for each business line is calculated by multiplying gross income (of that specific business line) by a factor (denoted beta) assigned to that business line.

   The total capital charge may be expressed as follows:

$$K_{TSA} = \frac{\sum_{t=1}^{3} \max\left[\sum (GI_{1-8} \times \beta_{1-8}), 0\right]}{3} \tag{1.5}$$

where

K_{TSA} is the capital charge under the Standardized Approach;
GI_{1-8} is the annual gross income in a given year, as defined above in the Basic Indicator Approach, for each of the eight business lines;
β_{1-8} is a fixed percentage, set by the Committee, relating the level of required capital to the level of the gross income for

Table 1.7 Business lines and beta factors for operational risk under Basel II.

Business lines	Beta factors (%)
Corporate finance ($\beta1$)	18
Trading and sales ($\beta2$)	18
Retail banking ($\beta3$)	12
Commercial banking ($\beta4$)	15
Payment and settlement ($\beta5$)	18
Agency services ($\beta6$)	15
Asset management ($\beta7$)	12
Retail brokerage ($\beta8$)	12

each of the eight business lines. The beta factors are detailed in Table 1.7.

3) Under the Advanced Measurement Approach (AMA), the regulatory capital requirement is given by the risk measure generated by the bank's internal operational risk measurement system, using quantitative and qualitative criteria subject to supervisory approval.

The approach or distributional assumptions used to generate the operational risk measure for regulatory capital purposes are not a priori specified, so banks can choose the best fit for their specific conditions. However, each bank must be able to demonstrate that its approach captures potentially severe "tail" loss events: Whatever approach is used, its operational risk measure must meet a soundness standard comparable to that of the internal ratings-based approach for credit risk (i.e., comparable to a 1 year holding period and a 99.9 percentile confidence interval).

Qualifying points of the methodology include the following:

• To calculate its regulatory capital requirement as the sum of expected loss (EL) and unexpected loss (UL);
• A bank's risk measurement system must be sufficiently "granular" to capture the major drivers of operational risk affecting the shape of the tail of the loss estimates;

- Include the use of internal data, relevant external data, scenario analysis, and factors reflecting the business environment and internal control systems;
- Have a credible, transparent, well-documented, and verifiable approach for weighting these fundamental elements in its overall operational risk measurement system.

Under the AMA, banks are allowed to consider the risk mitigating impact of insurance in the measures of operational risk used for regulatory minimum capital requirements, with a maximum of 20% of the total operational risk capital charge.

1.2.6 Basel III

Starting from June 2011, the BCBS introduced a comprehensive set of reform measures to strengthen the regulation, supervision, and risk management of the banking sector, and add a macroprudential overlay, aiming to improve the banking sector's ability to absorb financial and economic shocks.

The target of this reform is twofold: at bank level (microprudential, regulation), to raise the resilience of individual banking institutions to financial and economic stress; at system level (macroprudential regulation), to reduce systemwide risks across the banking sector and avoid the procyclical amplification of these risks.

The new framework includes a stricter definition of capital and a substantial strengthening of the counterparty credit risk framework.

The effects of the reconsidered loss-absorbing capacity of banks capital (see Table 1.8) nearly halved the previously considered effectiveness of the capital endowment as a shock absorber, as evaluated by the EBA quantitative impact study of 2011.

Apart from the definitions of capital and risk, the framework also includes an increase in minimum capital requirements, that is, from the 8% of RWA for total capital, 4.5% for Tier 1 and 3% for Common Equity are set to a new minimum of 6% for Tier 1 and 4.5% for Common Equity. On top of this, a capital

Table 1.8 Average estimated change in total capital ratio and RWA due to Basel III.

	Group 1 (%)	Group 2 (%)
Average Tier 1 capital ratio as of June 30, 2011	11.9	10.9
Average Tier 1 capital ratio under Basel III	6.7	7.4
Change in RWA due to Basel III	21.2	6.9

Source: EBA (2011).

conservation buffer of 2.5% is introduced for all banks, and an additional Common Equity Tier 1 (CET1) capital requirement ranging from 1 to 2.5% (depending on a bank's systemic importance) is set for the global systemically important financial institutions (SIFIs), to reflect the greater risks that they pose to the financial system.

So, the minimum capital requirement reaches 10.5%, for all banks but the SIFIs, for which an additional loss-absorbing capacity is required, ranging from 1 to 2.5% depending on the systemic role of the bank (Table 1.9), so that the total capital requirement ranges from 11.5 to 13%.

Another requirement is introduced with reference to liquidity, so that banks are required to have sufficient high-quality liquid assets to withstand a 30-day stressed funding scenario, a net stable funding ratio, and risk management; new references for capturing the risk of off-balance sheet exposures and securitization activities, managing risk concentrations, and other specific related problems.

The Basel III framework is planned to be fully in place in 2019.

The specific reference to global systemically important banks (G-SIBs) is based on the possible cross-border negative externalities posed by G-SIBS, which bring wider spillover risks the system must be protected from.

The moral hazard related to the implicit guarantee by the governments on too-big-to-fail financial institutions can lead to

Table 1.9 Basel III phase-in arrangements.

	2013 (%)	2014 (%)	2015 (%)	2016 (%)	2017 (%)	2018 (%)	2019 (%)
Minimum common equity capital ratio	3.5	4.0	4.5				4.5
Capital conservation buffer				0.625	1.25	1.875	2.5
Minimum common equity plus capital conservation buffer	3.5	4.0	4.5	5.125	5.75	6.375	7.0
Minimum Tier 1 capital	4.5	5.5	6				6
Minimum total capital		8					8
Minimum total capital plus conservation buffer		8		8.625	9.25	9.875	10.5
G-SIB additional CET1 loss-absorbing capacity				Gradual introduction			1–2.5
Liquidity coverage ratio—minimum requirement			60	70	80	90	100

Source: BIS.

suboptimal outcomes for the system, which, in consequence, has to include these issues in the policy design.

The main aim of the policies is to increase the loss-absorbing capacity of G-SIBs, so as to reduce its probability of failure.

The Basel Committee has developed a specific methodology based on indicators, for determining the systemic importance of G-SIBs, on the idea that it must be measured in terms of the impact that the bank failure can have on the global financial system and economy. While the crisis prevention is based on a reduction of the probability to default, the acknowledgment of

Table 1.10 Buckets and additional loss-absorbing capacity for G-SIBs.

Bucket	Score range	Minimum additional loss absorbency (common equity % of RWA)
5	D-	3.5
4	C–D	2.5
3	B–C	2
2	A–B	1.5
1	Cutoff point–A	1

the systemic importance is based on its impact on the system, and also on its loss given default.

The indicators used for assessing the systemic relevance reflect the size of banks, their interconnectedness, the lack of readily available substitutes or financial institution infrastructure for the services they provide, their global (cross-jurisdictional) activity, and their complexity.

The composite index is based on an equal weight to each of the five categories of indicators, each of which is normalized to a score of 1.

On the basis of their scores, G-SIBs are assigned to one of the four categories of systemic importance (Table 1.10), with varying levels of additional loss absorbency requirements.

1.3 Banking Risk Modeling and Stress Testing

The risk splitting introduced by the Basel II framework is operatively effective, even if it categorizes the risk in a strong division, while, in fact, these risks are cross-linked.

With reference to the regulation, some of these linkages are acknowledged. On the one side, the maturity transformation realized by banks implicitly includes interest rate risks, as the shorter and longer maturities can have different variations, which can induce significant risks of reduction, or even negative values, for the net interest income.

On the other side, counterparty credit risk is always present and has to be considered with reference to the trading activity, so it is strictly connected to market risk, as even traded assets are exposed to the risk that the counterparty could default before the final settlement of the transactions cash flows.

In addition, market and credit risk tend to be driven by the same economic factors. For example, both stock and bond values are sensitive to the macroeconomic environment changes. Hasan *et al.* (2009) proved that correlations between macroeconomic variables and asset prices, and prices of default-sensitive instruments are significant from both a statistical and an economic point of view. So, the same values are exposed to both counterparty and market risk variations, as a consequence of the same variables variations. One example can be in the effect of a worsening of the expected income of a bond issuer, which increases its default risk, thus inducing a reduction in market prices.

So, this separate estimation and aggregation, often used in industry and regulation, excludes nonlinear interactions and linkages, such as diversification benefits, and may lead to biases in the overall risk estimation.

A consistent approach must include the integrated evaluation of at least the credit and market risk, and include all sources of income and losses.

Some studies analyzed the two main risk sources from an integrated point of view, with reference to diversification, liquidity, and measurement.

Tang and Yan (2010) examine the impact of the interaction between market and default risk on corporate credit spreads. Using credit default swap (CDS) spreads, we find that average credit spreads decrease in GDP growth rate, but increase in GDP growth volatility. They also find that a major portion of individual credit spreads is accounted for by firm-level determinants of default risk, while macroeconomic variables are directly responsible for a lesser portion.

Drehmann *et al.* (2010) derived a consistent and comprehensive framework to measure the integrated impact of both risks and assess the integrated impact of credit and interest rate risk on banks' economic value and capital adequacy. They showed the importance of measuring and stress-testing the impact of credit and interest rate risk jointly, and that a deterioration in a bank's fundamentals can increase its funding costs, thereby further lowering its profitability in a potential vicious circle.

Breuer *et al.* (2010) showed that when financial positions depend simultaneously on both market risk and credit risk factors, an approximation of the portfolio value function separating the two risk components can lead to an incorrect assessment of true portfolio risk.

With reference to regulation, the integrated effects are one of the problems to be considered when setting the stress tests required for banks using the internal models approach. In this case, a rigorous and comprehensive stress-testing program must include the effects of the external shocks on the different sides of the banking activity.

Another possibility is in starting from the a posteriori statistical distribution of the actual results, which implicitly includes all the effects affecting banks' results. With appropriate time series and cross-sections analyses, it is possible to describe the actual banking expected losses distribution, and to evaluate the role of the different variables affecting it.

But even this approach suffers from several limits. One is in the difference between risk measurement and the actual expression of results variability: If we toss a coin three times, and register three "heads" results, this does not means that the coin is loaded, or that the fourth try will result in one more heads result. It is also one possible result of a standard coin tossing with actual probability of ½ for each possible result. The risk and results measuring are two different measurements that need different methods, even if each one is linked to the other.

Another important difficulty is due to the actual loss acknowledgment timing.

A liquidity problem for a loan debtor can result in its default, and the bank will register a loss hitting the income.

But if the bank allows for credit flexibility, raising the credit limit for the debtor, the firm can avoid the default, and the impact of this different strategy will just result in a higher risk weighting of the loan, and a higher correction in the expected losses. This results in a higher RWA value, but with a much smaller impact on income.

Thus, the actual results (losses) not only depend on the different risk profiles and diversification strategies chosen by each bank but also are influenced by the balance sheet policy.

In addition, the actual losses are almost always the result of a loan granted 1 or more years before the actual default, and it is not directly evident from the balance sheet which is the risk level of the exposures. Only a indirect evaluation can be done from the loan losses provisions, which refer to all the loans and exposures, and can be set with some subjectivity by the bank.

Finally, the a posteriori analysis of results distribution, even if it is not sufficient to assess each bank's actual risk level, is actually useful to picture the framework, evaluate the main influencing variables, and produce an adequate representation of the possible results of the banking activity.

1.4 Contagion

As already mentioned, banks have a key role in financing the real economy, sustaining the economic growth, and allowing a higher growth rate. In addition, not only do banks channel financial resources to the creditworthy firms, as do a substantial selection between the possible investments, but also the intermediation role of banks realizes a maturity transformation that increases the possibility of firms financing, which cannot be possible by the direct contact between investors and money suppliers. Thus, every bank default is a seriously important accident that induces significant effects on the whole economic system: losses for depositors, quite often consumer families, with social and economic disruptive effects; lack of confidence in banks, and thus a lower attitude to deposit money in banks and a lower capability of firms financing; fire selling of traded assets, with subsequent effects of markets instability, and so on. So, even when the bank default is not causing direct effects on the rest of the banking system, it still affects the system owing to its indirect effects.

Evidently, the larger the bank, the larger the effects on the rest of the system, but the worst case after a bank default is due to contagion.

Similar to what happens for some diseases, where the illness of one person can spread to more people, the default of a bank can be the cause of other banks' defaults, and this propagation can continue in a domino effect, transforming a bank crisis into a systemic crisis. If the default of one bank is already a significant problem for its impacts in the real economy, systemic crises have disruptive effects on the economic system and public finances, often affecting the real economy to a large extent, not only in its values for the years of crises, but also changing the trend for subsequent years.

The recent financial crisis demonstrated what the consequences of a banking crisis can be. But this is just the last example: Laeven and Valencia (2013) reported 147 banking crises episodes over 1970–2011, and a median value of 23.2% of output loss across all episodes reported. Figure 1.4 reports some examples of the effects on GDP of banking and financial crises.

The knowledge of the disruptive effects a systemic crisis can have on the economic system induced the regulators and supervisors to analyze in depth the mechanisms possibly causing a systemic crisis, and in particular bank defaults and contagion.

In fact, contagion channels and mechanisms are multiple, and we already mentioned the loss of confidence and the destabilizing effects of fire selling; others are related to the derivative markets, to the interrelationship with public finances, but the most significant and direct effect is due to interbank lending.

As considered with reference to the counterparty risk, even when the counterparty is a bank, the risk that a loan will not be repaid is always present. Thus, the default of a bank normally implies that the interbank loans are not, or not fully, paid back, so the creditor bank suffers from direct losses. In addition, as the interbank lending is part of the liquidity management, even the indirect effects, such as interbank lending freezing, restoring liquidity equilibriums, and the subsequent effects on the other

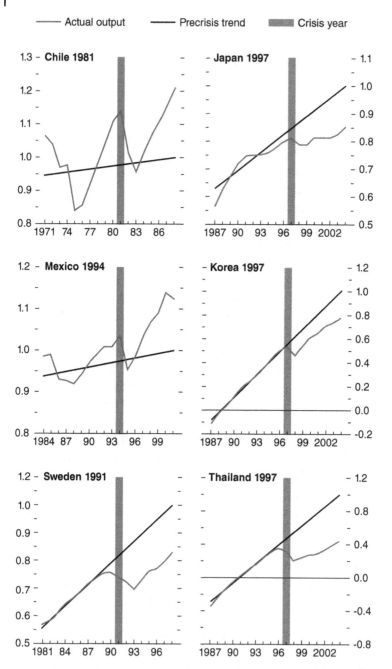

Figure 1.4 Medium-term output per capita after financial crises: case studies. *Source:* IMF (2009). Reproduced with permission of IMF.

layers of the banking activity, result in higher costs or in lower income for the affected bank.

And when these losses are important, or when the bank is already weakened by other problems, the creditor bank can default itself, causing more interbank losses to other banks, possibly summing up to the previous default losses, and so on. This chain reaction is one of the worst fears of the supervision institutions, the costs for stopping the mechanism being huge, and in some cases above the national government capabilities.

One important piece of evidence, similar to what happens for diseases, is that the earlier the intervention, the lower the impact of the crisis and the costs of stopping the domino effects. So, quite often possible bank defaults are stopped with early interventions, as the default can be the lighter of a burning disaster. But if the crises have important costs, even the interventions for preventing it are costly, so it is important to have a correct perception of which crises can introduce serious risks to the system, in order to have more effectiveness and efficiency in using the tools designed to maintain financial stability.

The first evidence is in the role of the size of the defaulting bank. Large banks are so important that they can induce a systemic crisis just for the size of the losses induced to the counterparts. Quite often, the larger banks are qualified as "too big to fail," as no government would be able to withstand the effects of their default, so early interventions are always put in place to avoid it.

This also induced a number of distortions related to possible moral hazards. In fact, some rating agencies when evaluating the soundness and safeness of banks also include in their evaluations the attitude of governments to intervene to avoid a possible default. So, the "too big to fail" or "systemically important" banks benefit from a higher rating, because of better market conditions due to this implicit guarantee. But in some cases it can also be exploited in moral hazard terms by management, and taking more risks than appropriate, knowing that if the risk outcome is a higher profit, this will benefit for the management (and shareholders), while if the risk outcome is a loss, the government (taxpayers) will pay to cover the losses. Evidently, some mechanisms have been adopted in many countries for correcting this strain.

Another important evaluation is detecting the kind of shock causing the bank default. The main distinction is between the so-called idiosyncratic shocks—risks specific to the affected bank—and systemic shocks, variations in the economic framework affecting all banks, or a large part of banks in the system.

Examples of the first case can be the crisis of a large firm that is a big borrower from a bank, a local housing or specific crisis for a highly specialized bank, or the effects of fraud. In these cases, when the shock is seriously affecting only the defaulted bank, and when the defaulted bank is small, the counterpart banks can absorb the interbank losses. Instead, when the shock is due to a widespread economic crisis, financial market instability, or any other problem affecting the whole system, as many banks are already weakened, they can be seriously hit by more losses coming from interbank defaults, and so are more likely to default and be involved in domino effects.

Castiglionesi (2007) investigated the role of central banks in preventing and avoiding financial contagion, and found that the liquidity reserve role is fundamental in facing adverse shocks that could cause contagious crises.

It is also important to consider the central bank role as a lender of last resort that can also play the role of money centre for interbank lending during the banking crises, as banks have a lower confidence in the interbank lending and the use of the central bank as counterpart can restore confidence, make the interbank market work, and avoid the liquidity shortage that characterizes systemic crises.

Co-Pierre (2013) proved that the liquidity provision by the central bank results in a significant stabilizing effect. However, its effects are only significant above a certain threshold; this stabilizing effect is nonlinear, so that even slight changes in the collateral requirements can have significant stabilizing effects if performed around the critical value; and the precise value of the threshold depends on the specific parameterization and network structure.

1.5   System Modeling

After considering the different problems that banks have to face and the risks they have to manage, the analysis can be aimed at studying the banking system as a whole.

The main components characterizing each banking system are the distribution of the banks' values, the direct linkages of interbank lending, and the specific framework they share, in terms of loans market, national economic policy, and in particular banking and financial policy and supervision, and the direct linkages. In terms of modeling, this is translated in contagion transmission channels and correlation among bank results.

The distribution of banks, values is the way of mapping dimensions and roles, and evaluating the system structure.

Domestic banking systems can be characterized by a large number of small banks, or concentrated into a small number of larger banks. Quite often, the different dimension classes correspond to different roles in the system. Smaller banks are typically characterized by relationship lending, direct deposits funding, and high assets shares invested in loans, while large banks tend to have a more structured activity, including trading, corporate finance, and merchant banking.

The definition of a banking system is typically based on the homogeneity and on the linkages among the considered banks.

In this sense, quite often the analyses are referred to national banking sectors, as the homogeneity of laws and reporting standards and the common counterparts, both on the loans and on the policy and supervision.

The national policies the banks are exposed to refer not only to specific banking regulation and interventions but also to all the economic framework. The government countercyclical actions for smoothing the business cycles, support for specific economic sectors or regions, and the public finances equilibrium are only some of the examples of the policies directly affecting the banking activities and results.

The public finances are particularly relevant with reference to the banking system, as the channels linking public finances and banks are multiple, both sectors being exposed to the interest rate and money quantity control by the central bank, the country risk affecting both ratings of sovereigns and banks, the role of banks, and so on.

But the direct linkages between banks and public finances are due to the exposure of banks as sovereign bonds holders, and of the government as an implicit guarantee of rescuing the bank in case of default.

On the exposure of banks in sovereign bonds, it is worth noting that the large size of a banking group makes it an important market maker. This is particularly significant for countries with a large public debt, thus requiring continuous support for the bonds issues, giving these banks a strong position in its relationship with the government. On the other hand, banks are exposed to possible losses as a consequence of the public finances weakening.

The government's implicit guarantee is a consequence of the evidences of the banking crisis effects: When contagion spreads, the cost of nonintervention or of a delayed intervention can be thousands of times higher than the early intervention—preserving the agents' confidence and avoiding the nervous and often irrational sudden changes in the financial sector behavior limit the disruptive effects of the crisis spreading over the economic system. But this contingent liability and the actual costs of rescuing the banks can substantially hit the public finances equilibriums. In a weak phase of the business cycle, this mutual support, of banks sustaining sovereign bond prices by buying and holding important shares of public debt, and of governments guaranteeing for banks rescue, can be a tough exercise.

The IMF Committee on the global financial system (2011) described the risk of lower profitability for banks as a consequence of public finances weakening mainly passing through four channels, as explained by and shown in Figure 1.5:

- A fall in the value of the government bonds held by banks, weakening banks' sovereign portfolios.
- *An increase in banks' funding costs:* A deterioration in a sovereign's creditworthiness reduces the value of the collateral

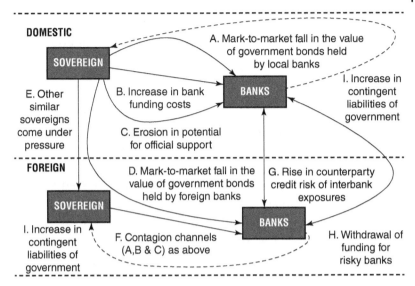

Figure 1.5 Spillover effects from sovereign to banks and from banks to sovereigns. *Source:* Laeven, 2010. Reproduced with permission of IMF.

that banks can use for wholesale funding and to obtain liquidity from the central bank.

- *Erosion of the potential for official support:* The affected sovereign's ability to backstop the financial system comes into further doubt, as rising funding costs increase the magnitude and likelihood of interventions in banks.
- Sovereign downgrades generally cause lower ratings for domestic banks, increasing their wholesale funding costs, and potentially impairing their market access.

In addition, the role of banks in funding firms and supporting the economic growth makes the domestic banks an important interlocutor for any government, and the banking sector one fundamental actor in implementing and realizing the economic policy.

With reference to the modeling, it is fundamental to highlight that many risk factors affect all banks in the system, even if with differentiated impacts.

The first common exposure is to the national business cycle, which affects firms' economic framework, and ultimately their capability to pay back bank loans.

Because all firms (and customers) are affected by the same business cycle, the results of counterpart risks for different banks are correlated. This effect is enhanced by the possibility, often realized, that the same firm is actually financed by more banks, as, in this case, the risk is shared by all the concerned banks. Other risk sources shared by the banks in the same system are related to interest rates, domestic stock exchange values and trends, policy interventions affecting propensity to save or consume, and so on.

With reference to linkages, a system can be characterized by its interbank direct exposures, so that a group of banks can be reasonably defined as a system if their linkages are mainly held to other banks in the system. In this way it is possible to reasonably approximate the possible states of the system, and include the main determinants of its possible instability. The same problem can be referred to a system represented by only a sample of banks that can correctly approximate the actual system as soon as the banks in the sample banks are actually representative of the system distribution.

For large banking groups, the framework is significantly different. On the one hand, large banks are typically operating in more countries, and hence are exposed to more risk sources, while on the other hand, diversifying risks are coming from different countries. This means that when modeling their risks and behavior either the correlation is referred to international business cycles or its results are represented as the weighted average of the economic framework for each considered country.

The second difference is that quite often, due to the inter-mediated volumes, large banking groups are exposed among them, in an upper layer of the banking network. In this case, there is an actual homogeneity between the large and internationally active banks and banking groups (so-called "group 1" banks), so that it can be modeled as a system even if there is no national communality.

What is in fact fundamental is to represent, as correctly as possible, the actual system behavior and dynamics.

2

Simulation Models

Most of the literature on banking is based on balance sheet and market data. This approach has several advantages, as it looks at the banking activity as a whole, and even when the analysis is focused on specific aspects of its activity, it is always based on real banks actual equilibriums. In fact, banks have to meet complex equilibriums on different aspects, such as loan pricing, liquidity, risks, maturity matching, and income. The use of data coming from real banks ensures that these equilibriums are respected; as the considered banks exist and act in the real-world economy and market, all data considered are always actually possible states of the banking activity.

Regression analyses focused on activity, lending selection, scale economies, cost and profit functions, and pricing, to name but a few examples, derived from this approach.

However, in some cases this is not possible, since real-world balance sheet and market values do not provide adequate information, such as when the events are really rare, or the available episodes are not comparable due to regulation changes, different accounting frameworks, or other. This problem becomes important when dealing with rare but important problems (the so-called black swans) that cannot be neglected for their importance, but are so unlikely to happen that only a small number of observations can support the economic analysis. In these cases, one can employ a theoretical approach to build a reference framework that correctly represents the

Banking Systems Simulation: Theory, Practice, and Application of Modeling Shocks, Losses, and Contagion, First Edition. Stefano Zedda.
© 2017 John Wiley & Sons, Ltd. Published 2017 by John Wiley & Sons, Ltd.

various aspects of the activity, and then use this "laboratory" to analyze the phenomenon and its expected effects.

This different approach induces fundamental differences in both the representation and the use of results.

With reference to representation, the model must include all the main determinants and mechanisms involved, since, in contrast to what happens when dealing with real data, the implicit equilibriums of the banking activity are not necessarily preserved.

In this representation, all the information is stored in the model functions and input values, so it only depends on the model to provide for an adequate representation of the actual processes.

With reference to the use of results, it is important to keep in mind that the model is structured for correctly representing some aspects of the activity, and keep the main references for the other aspects, but it does not necessarily give a good representation of the aspects that are not specifically considered as an aim of the representation. So if, as an example, the aim is a representation of the possible losses for the bank or for the banking system, nothing ensures that the model results give an appropriate representation of the liquidity needs or of the possible business model adaptations.

The second fundamental reference to the use of results is that all results are only determined by the model values and functions. This makes a fundamental difference from the real data (regression analyses). While dealing with real data, quite often the aim is to find or better describe some mechanism determining the results of the activity. When dealing with models that do not interpret the banking activity, but try to reproduce it, all the mechanisms are due to the modeling, so the focus of these exercises can be on more complete representation of the possible outcomes of some process or on what-if analyses, which can explore the effects of changing the input values.

The mathematical representation of complex structures is the basis of engineering models. Such a mathematical approach has also been applied in economics, but only in the last few decades has it been employed in business studies.

With reference to banking, this approach has revealed its importance when dealing with banking crises, including defaults and contagion. In fact, bank defaults are hopefully not so frequent in the real world, but their effects are devastating in terms of consequences on other banks, financial markets, public finances, and the real economy.

Furthermore, banks are in some cases extremely large firms (in some cases larger than their home country GDP), with multinational activity, and the banking sector has the highest leverage, in terms of assets per unit of equity, with regard to all other sectors of the economy.

This means that only a small part of the assets is backed with capital, while the largest part of the assets is covered by debts, as deposits, issued bonds, and interbank loans. So, when a shock hits a bank, the loss-absorbing capacity of the capital buffer can be insufficient, so the bank defaults, causing serious problems to the debt holders, depositors, bond holders, and interbank counterparts. The actual effects can be so important as to destabilize one (or more) country's financial and real economy.

This small capital base and large size mean that banks can be represented as tall buildings with a small base that can collapse on the neighboring buildings, crashing or damaging them and substantially affecting the city life.

This has opened several different streams of research: What can happen to the rest of the system when a bank defaults? Why do banks default? And what if more banks default at the same time?

Besides looking for answers to these questions, by means of the analysis of the variables that can lead to bank defaults, and of the linkages and effects of bank defaults on the system, the debate proceeded to a deeper question: What can happen to the system if the subjacent variables are so negative to make some banks default?

This is a more complex scenario, where all banks in the system are affected by the same negative subjacent variables, and the crash of one or more banks happens in an already-weakened system, like an earthquake that affects all buildings in a city, making some of them crash on other buildings already weakened by the same earthquake.

In these cases a domino effect is likely to develop, with huge effects on the whole economic system.

Luckily, we do not have many cases of financial domino effects, and therefore not enough data are available for analyzing in detail the mechanisms and deriving adequate conclusions on what could happen in similar cases.

So, to understand the possible effects of a banking crisis, it is necessary to split the process into its basic components, and analyze them one at a time. Such basic components can be summarized as the variables affecting banking results, the safeness of banks as single entity, their interconnections with the other banks, and the contagion mechanisms. We can then put all the components together to develop a reliable simulation of the process, and derive some references on what could happen and how to reduce the probabilities of banking crises and contagion.

Even when representing the banking system as a simple model, we have to deal with several components and effects. The first one is the risk sources, which are affected by (at least) some autonomous characteristics and some common factors (Figure 2.1).

The effect of these risk sources is to create a significant variability in results, in particular due to credit losses, but also due to market or operational losses. In the credit activity,

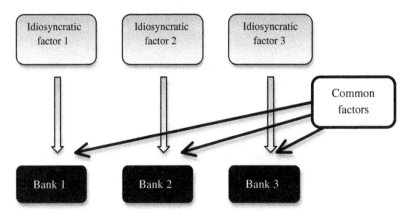

Figure 2.1 Banking system idiosyncratic and common factors.

the risk that some loan is not paid back, that some bond is sold at a lower price than it was bought, or that some problem happens in the activity is not only known but also expected, so the income equilibriums are based on its expected values. The presence of the risk sources implies that in some years results are higher than expected (lower losses), and in some other years results are lower than expected, i.e. losses are higher than expected.

The effect of the loss variability is, almost partially, absorbed by capital, which can grow when results are more positive than expected, and is reduced when losses are higher than expected.

While in the first case there is no problem, as profit can be stored as higher capital or distributed among shareholders as dividend, in the second case, when losses are higher than expected, the loss-absorbing capacity of capital is limited: If the capital falls below zero, the bank defaults, unless some sort of recapitalization is implemented.

Besides the risk sources, the second component that we must consider when developing a bank system model is the capitalization level.

The third component is related to the bank interconnections that can cause banks to fail if a counterpart in interbank loans defaults, or for its indirect effects (fire selling, price variations, derivatives, etc.).

This third component, known as contagion or the domino effect, is hopefully the least significant in terms of probability, but also the most important in terms of its impact.

In the following sections, this complex mechanism including risk sources, default modeling, and the contagion mechanism is analyzed, step-by-step, also following the literature developments.

2.1 Simulating Shocks: Idiosyncratic Shocks, or Exogenous Failure of Individual Banks

The first and simplest exercise, designed to analyze banking system stability and contagion simulation, consists in assessing

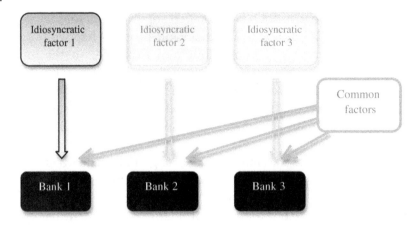

Figure 2.2 Banking simulation: idiosyncratic bank failure.

the effects of an unprecedented bank failure. The basic idea is to verify if the default of one bank can, *ceteris paribus*, induce by contagion the default of other banks.

Papers by Humphrey (1986), Angelini *et al.* (1996), Sheldon and Maurer (1998), Furfine (2003), Degryse and Nguyen (2007), Wells (2004), and Upper and Worms (2004) analyzed this case, verifying what the consequences can be (Figure 2.2).

Even though this is an artificial scenario, where no motivation is given on the initial default and all other banks are supposed to receive no other shocks besides contagion losses, this exercise led to some important conclusions.

Having a simple and clear initial shock is in fact a way of simplifying the framework and focusing attention on contagion mechanisms.

A bank's balance sheet can be simplified as shown in Table 2.1.

In this simple scenario, one bank is considered to fail for an exogenous reason, possibly derived from the specific bank activity, and as a consequence the defaulting bank does not pay back the interbank debts.

More formally, in the case of distress of bank i, the bank j, creditor for the amount IB_{ij} will experience a loss given by

$$L_j = IB_{ij} \times LGD_i \tag{2.1}$$

Table 2.1 Simplified representation of a bank's balance sheet.

Assets	Liabilities and equity
Loans $\displaystyle\sum_k A_{ik}$	Deposits
Traded assets	Interbank debts $\displaystyle\sum_j IB_{ij}$
Sovereign bonds $\displaystyle\sum_c SB_{ic}$	Other liabilities
Interbank credits $\displaystyle\sum_j IB_{ji}$	Equity K_i
Other assets	Profit/losses L_i

where LGD_i is the interbank loss given default (LGD) of the distressed bank i, so the amount of its interbank exposure that, as a consequence of the default of the debtor bank, will not be recovered.

This contagion channel is in fact of fundamental importance even in more complex frameworks, where other effects are also considered.

The basic default propagation mechanism is rooted in the fact that, if the losses suffered by contagion are higher than the capital coverage, the affected bank will default itself, thus becoming the infectious vehicle for the next cycle of contagion. This chain reaction goes on until either the shock reaches robustly capitalized banks, so that no more banks default, or the entire system defaults.

Two characteristics of the system are fundamental for this process. The first one is the step effect, which entails that even one single dollar difference can make a bank default, thereby changing the subsequent steps of the process. The second important characteristic is in the values determining the possible default. On the one hand, we have the losses dimension, first due to the interbank exposures value (and LGD rate), which determines the shocking component dimension, and, on the other hand, we have the capitalization level of banks, as capital is the shock-absorbing component, which (when sufficient) neutralizes the impact of the incoming shock.

The step effect cannot be underestimated when simulating the system, as the consequences of a default are always significant even if the default is for a really small amount. This trigger effect has to be included in modeling even if it induces difficulties on the mathematical representation of the process.

With reference to the values, the capital level can be easily obtained from any bank balance sheet, even though the actual shock-absorbing capacity of the diverse components of bank capital is subject to important debate and analyses.

The second fundamental value is given by the interbank exposures.

The main problem when dealing with the interbank exposures' data is due to the unavailability of the exposures' data by single counterpart. This is due to two main problems. The first refers to the accounting framework, as banks only report the total value of exposures, and not the single values of the exposure to each counterpart bank.

The second is due to the exposures' high variability, so that even if the values were known at the balance sheet closure date, this value would only be a reference, as the values change daily, and it is not possible to have a precise quantification of the exposures at a different date.

Deeper analyses and references on this estimation problem are provided in Section 2.9.1.

The third fundamental evaluation refers to the loss given default, or LGD.

In fact, the consequence of a default is not in the loss of the entire exposed amount, since typically a significant portion of the value can be recovered, and only a smaller fraction is lost.

What is less evident is that the loss suffered by the creditor bank is not only due to the direct unrecovered value but also due to the need for rebalancing liquidity and all the financial equilibriums, which often force the bank to sell some assets and revise the risk coverage. The need for a quick rebalancing often changes the assets sale into fire selling, so that more losses, indirectly due to the interbank default, need to be taken into account. This fire selling effect is worsened when the initial bank default is important, so many banks have a similar problem and the impact on markets is of a simultaneous fire selling, leading to higher impact on selling values.

A detailed analysis of the LGD evaluation is reported in Section 2.10.

On the modeling side, even the simple approach of only considering the consequences of one unprecedented default already highlights the main issue: Because of the contagion step effect, it is not possible to derive solutions in closed form (i.e., by solving a system of equations) and therefore an analytic approach is not feasible.

Moving on to a formal representation of the process, we need to include the two impacting variables: the amount of the interbank losses and the capital level of the infected bank.

Following Merton's (1974) approach, bank j defaults when its losses L_j are higher than the capital endowment K_j:

$$L_j > K_j \tag{2.2}$$

So, when the value of its exposure to the initially defaulted bank i, IB_{ij} times the actual loss given default LGD_i is higher than the capital:

$$IB_{ij} \times LGD_i > K_j \tag{2.3}$$

This mechanism includes a step effect, triggered by the level of losses, that induce deeply different consequences, depending on the default or not of bank j. This nonsmoothness is actually complex to deal with if using an analytical approach, but quite simple if tackled with computational and simulation methods. This is even more significant when considering more complex mechanisms including correlation or cross-linkages.

In fact, even when dealing with a single exogenous failure, more contagion channels or other effects can be considered, or some evolution in modeling for including more complexity, as many recent papers did.

The interbank exposure default can also induce more defaults, when the loss suffered by the creditor bank reduces the value of its assets so that the net capital value becomes negative. In such a case, the creditor bank defaults by contagion and its interbank exposures will affect the other banks, adding up to the initial bank default effects.

More sophisticated approaches include taking into account endogenous or exogenous LGD, netting or not the interbank mutual exposures, the inclusion of market-induced instability, the endogenous modeling of fire selling and other bankruptcy costs, and different methods for estimating the interbank matrix.

Further evolutions of the model can include considering different seniorities for nonbank liabilities, including derivatives exposure as a channel of contagion or credit risk coverage, limiting (or not) the analysis to domestic exposures, limiting (or not) the effects of contagion to overnight or small residual maturity exposures, and including credit risk transfer mechanisms or the presence of safety nets.

This is just a list of some references developed in the literature, but the continuously changing regulation and banking business models will drive the modeling to represent as effectively as possible the actual framework, including new mechanisms.

2.2 Simulating Shocks: Stress Testing

After evaluating what happens to the system when one bank fails, the analysis can be developed to analyze the causes that can bring the bank to default, which are the variables affecting bank results, and assessing what happens to the bank if one risk source hits the bank. These exercises, known as stress tests, had an important role in the recent banking supervision, as it allows for testing the actual bank resilience to possible economic and financial crises.

There are two approaches to designing the stress testing scenarios.

The first approach, in a theoretical exercise, specifies the stress testing scenarios based on the statistical properties of the shock variables in the model, so as to compute which effects are to be expected, for example, in the worst 1% of cases (the worst year in one century).

The second approach is aimed at reproducing the initial shocks that induced the actual crises, so a historical approach, in order to compare what actually happened in the considered episodes to the estimated effects of the same shocks in the tested framework. The most cited references are to the Asian crisis in 1997, the LTCM crisis (1998), the 2007–2008 subprime and global crisis, and the 2011 sovereign crisis.

One important study in this respect is detailed in the paper by Laeven and Valencia (2013), which provides a list of the crises that occurred in previous decades.

The two kinds of stress tests have different aims. The test for the sensitivity to risk sources is aimed at evaluating the effects on bank results of the framework changes in a more "normal" and typical situation, while the second kind tries to verify if the (hopefully) less frequent financial storms, where many variables can reach unexpected values, can actually destabilize the bank.

The two approaches are complementary, and combining the two, it is possible to provide a more complete picture of the financial system vulnerability, from both statistical and historical points of view. The first approach can be a more complete exercise, evaluating many possible combinations of different stress conditions. The second approach, instead, in reproducing actual crises, reflects more realistic scenarios and allows a comparative analysis over time.

With reference to single banks, the BCBS references suggest that banks should subject their portfolios to a series of simulated stress scenarios of the two different kinds, reproducing the recent periods of significant turbulence and assessing the sensitivity to variations in the risk sources and correlations, and provide supervisory authorities with the results.

A more sophisticated analysis includes the identification, in a forward-looking approach, of the most adverse scenarios for the actual bank portfolio, based on possible specific crises (jumps in oil prices, problems in a specific region, etc.), so as to prove the banks' capabilities to identify the actual risk and the banks' resilience to these possible crises.

In 2011, the European Banking Authority (EBA) started a systematic stress test program to assess the resilience of the main European banks to adverse market developments and to contribute to the overall assessment of systemic risk in the EU financial system. These stress tests are conducted on both trading and banking book assets (including off-balance sheet exposures), and focused on assessing credit and market risks and understanding specific weaknesses in the solvency of banks.

While the main source of possible losses is identified in credit risk, market risk was also included in the exercise, with a specific focus on the sovereign risk exposures, including its effects on the evolution of the banks' funding costs

The same approach applied to single banks can be used for stress-testing the whole system.

2.3 Simulating Shocks: Systematic Common Shocks

Many of the studies cited in the previous sections are based on the simulation of a single shock: a single bank default. However, this simplified framework is reasonable when studying the contagion effect, but if the goal is to evaluate the system stability as a whole and the effects on the system of a realistic shock, this approach is too limited.

A more direct way of representing the role of common risk sources is to explicitly include one common factor in the model and to estimate the exposure of each bank to it (Figure 2.3).

Since the exposure to a common risk source is one of the most important systemic risk factors, some attempts to analyze the effects included in this approach were developed in the recent literature.

One interesting example is in Paltalidis *et al.* (2015), where the risk sources are considered on the basis of the main assets classes in a bank's balance sheet. In this work, since the largest part of the bank's investments is in customer loans, sovereign bonds, and interbank loans, the shocks are simulated as coming from these three different sources.

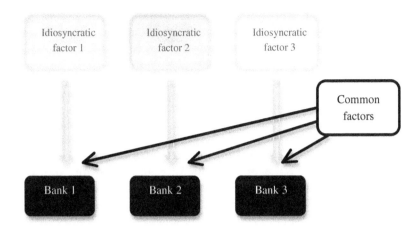

Figure 2.3 Banking system simulation: common shock.

The three risk sources are considered as totally separated and independent so that the shocks, which are induced as a 10% reduction in value for one asset class at a time, do not have any direct effect on other assets. Thus, the impact on each bank of the shock on a given asset only depends on its exposure to it, and on its consequences, such as defaults, fire selling, and so on.

Other possible approaches include to consider, as risk source, some macro vaiables. For example, the common risk sources can be the GDP variations, the interest rate variations, the exchange rate variations, the commodity price variations, and so on.

After evaluating which macrovariables are to be considered, the modeling must include the possible correlation between the risk factors and an estimate of each bank's exposure to each risk source, in order to have a reliable simulation of the actual shocks that the system is exposed to.

A typical representation of the risks each bank is exposed to relies on a split into two main components: the first, idiosyncratic component, is specific for the single bank and includes the particular exposures of that bank, and the second includes instead the common risk sources, i.e. the risks affecting all banks.

2.4 Simulating Shocks: Common Shocks

The external shocks that a system is exposed to are multiple, are diversified both in source and in strength, and can affect banks in different ways. Moreover, when considering common shocks, it is important to keep the focus on the correlation or on the interaction between the idiosyncratic component and the common factor(s), which affects each bank in a different way, possibly enhancing or absorbing the external shock.

The comparison between idiosyncratic and common shock effects is important for a more realistic representation of the framework. A test in this line is developed in Steinbacher *et al.* (2016), where the comparison of the effects of idiosyncratic and common shocks shows that while idiosyncratic shocks are the main source of single bank risk, they are not strong enough to destabilize the system. In fact, when the shock is only hitting one bank, the remaining system resilience is able to absorb the interbank losses; a common shock of even moderate magnitude can affect and weaken all banks at the same time and can bring a huge contagion potential.

So, in order to design a more realistic experiment, it is fundamental to include common shocks in the model, so that the actual structure of the shocks that banks and banking systems are exposed to can be represented as correctly as possible.

The approach described in the previous section, based on a single shock hitting the system, is no more adequate when dealing with common or correlated shocks. In mathematical terms, we must move from one binary variable (default or no default) to several continuous and correlated variables.

This not only induces a deeply different treatment of the variables but also adds a further step in the process.

The new step refers to the modeling of the reason why banks default. In the previous section, it was simply assumed, without any further motivation. In the framework considered from this section on, the model starts from one previous step, focusing on the variables that induce the uncertainty in results, and possibly causes the default.

When the inputs of a system are uncertain, the evaluation of its performances is necessarily uncertain. As a consequence, the best way for representing the results of any system based on uncertain inputs is a probability distribution.

While the result of a single simulation of a system is typically a single value ("if bank *a* defaults, bank *b* will default by contagion"), the result of a simulation based on uncertain inputs is a quantified probability ("if bank *a* defaults, there is a 35% probability that bank *b* will default by contagion"). This kind of analysis that includes both the uncertainty of the default of bank *a* and the uncertainty of the balance sheet results of bank *b* before including contagion losses is typically much more useful to decision-makers.

The complexity of this model of evolution lies in its mathematical complexity, which is due to the nature of the variables (continuous instead of binary) and due to the interaction between variables (correlation).

An example can possibly clarify the problem: If we have to model a simple system of, say, five banks, each one exposed to uncertain shocks, how can we test for the stability of the system?

On the one hand, as the reactions of the system are different for different shocks, there is not a single test that can fully describe the system stability.

On the other hand, it is not possible to consider all the possible values, as the set of inputs for this process is a five-dimensional continuous space.

An effective and actually used approach is somewhere between the two: explore the set of possible inputs without actually considering *all* possible values, but considering an adequate sample of *many* possible values, in the so-called Monte Carlo method.

2.4.1 The Monte Carlo Method

The Monte Carlo method is a technique used to understand the impact of risk and uncertainty, and it can be used to solve mathematical or statistical problems. The method is particularly useful when it is difficult or impossible to use other mathematical approaches. The Monte Carlo method relies on the substitution of any factor that has inherent uncertainty with a variable characterized by the appropriate probability distribution.

Coupling the Monte Carlo method with a simulation model is thus an important extension of the "what-if" approach described in the previous sections. When we introduce a variable that represents the risk source, rather than just considering a single shock, it is obviously much more demanding to determine what the distribution of possible results can be. Moreover, if several risk sources, possibly correlated, are considered, the Monte Carlo simulation ends up being the only chance to analyze the framework in terms of its probability distribution.

Like most gambling games in the casinos in Monte Carlo, the method is based on a random drawing, even though it applies not to cards or dice, but to some input variables. By drawing as many times as possible, and computing for each value the resulting effects on the considered system, one can compute the empirical distribution of the systemic effects.

The approach can be presented by a simple example. It is known that throwing dice can only give values from 1 to 6, each value with a probability of 1/6 (if dice are regular). It is also possible to know the distribution of the sum of three dice rolled together, and the probability that such a sum exceeds a preset value, say 15.

The problem can be solved analytically, using the classical rules of probability theory, but it is also possible to approach the problem in a numerical way, using the Monte Carlo simulation. The method in this case consists of the following steps:

1) Generate three random integer numbers, uniformly distributed from 1 to 6.
2) Sum up the three numbers.
3) Repeat the two previous steps a large number of times.
4) Count for the occurrence of each possible result (3–18) for computing the probability distribution of results, and of the values above 15.

The example shows how the method works and its main characteristics. The first evidence is that the input variable must be known not only in its possible values, but also in the probability distribution of its values. In the example above, if the input variable generated is not an integer, or if the random number generator does not generate the numbers from 1 to 6 with the same probability, the resulting probability distribution of results will be biased.

The same attention must be focused on the second step, in which the considered system behavior is simulated. In the example, it simply consists of summing up the three values, but typically the method is applied to complex systems, where the analytical tools are not applicable due to the systems' complexity.

With reference to the third step, it is evident that the larger the number of iterations, the more reliable the representation of the resulting probability distribution.

Figures 2.4 and 2.5 show how the number of iterations impacts the results. While, in Figure 2.4, obtained after 100 iterations, the probability distribution of results is obviously quite different

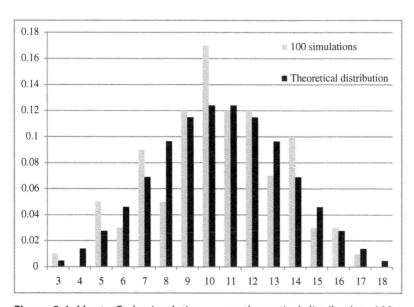

Figure 2.4 Monte Carlo simulation versus theoretical distribution: 100 iterations.

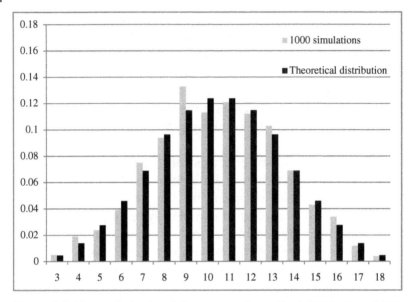

Figure 2.5 Monte Carlo simulation versus theoretical distribution: 1000 iterations.

from the theoretical distribution, the differences between the two are significantly reduced by increasing the number of iterations, as shown in Figure 2.5 just increasing it to 1000 iterations.

Due to this characteristic, Monte Carlo simulations are typically performed with a very large number of iterations, from tens of thousands to millions of cases, depending on the system complexity.

When dealing with more complex systems, the method follows exactly the same steps, apart from the obvious generalization:

1) Generate the input variables by employing their specific random distribution.
2) Process the input values via the simulation model.
3) Repeat the two previous steps a large number of times.
4) Count the occurrence of each possible result and compute the probability distribution of results.

The results of Monte Carlo simulations are a large number of separate and independent values, each one representing a possible outcome for the system, that is, one possible state that the system may reach. These independent system realizations are then assembled, so that the outputs, instead of single values, are probability distributions for the output variables.

Presenting all the characteristics and features of the Monte Carlo approach is beyond the scope of this book. More detailed references can be found in Kroese *et al.* (2011).

2.4.2 Monte Carlo-Based Simulation Models

As suggested in the previous sections, a reliable banking system simulation must include more risk sources, like the idiosyncratic components, affecting each bank differently, possibly correlated among them, common shocks, affecting all banks on the basis of their exposure to the common factor(s), and the step effect due to the possible default.

An analytical model of this framework, which includes the probability distribution of idiosyncratic shocks, common shocks, correlation among risk sources or common factors, exposures, and step effects, is neither actually known nor realistically feasible. Instead, the Monte Carlo approach allows us to break the problem into simpler components, for each of which we can then develop a simulation model. Finally, we combine all results into a final, complete model, in order to perform the iterations needed and derive the requested probability distribution for the system. Many recent papers simulated common shocks via Monte Carlo simulation, starting from Elsinger *et al.* (2006a, 2006b).

The Monte Carlo simulation process is normally carried out one line at a time and repeated as many times as needed, and the result of this kind of simulation is typically represented by a matrix where rows represent simulations and columns are banks. Depending on the simulation needs, the matrix elements can contain losses, excess losses, covered deposits, or a binary variable just signaling the default or no default.

For a panel of s simulations for i banks, starting from a random generated matrix of values R, we will have

$$R_{(sxi)} = \begin{pmatrix} R_{1,1} & R_{1,2} & \cdots & R_{1,i} \\ R_{2,1} & R_{2,2} & \cdots & R_{2,i} \\ \vdots & \vdots & \ddots & \vdots \\ R_{s,1} & R_{s,2} & \cdots & R_{s,i} \end{pmatrix}$$

An appropriate function $f(R)$ will transform the random values R into simulated losses L:

$$L_{s,i} = f(R_{s,i})$$

In this example the random generated variables are one for each bank for each line, but if the model includes common shocks, they can be included as one (or more) additional variable, which will be combined for obtaining the simulated losses.

The models developed for obtaining simulated losses are described in Sections 2.3 and 2.6.

In the resulting matrix L, each element represents the initial credit losses for one bank in one simulation. Typically losses are correlated among banks (columns), while different simulations (rows) are uncorrelated.

The net losses NL are obtained by subtracting the capital value to each element. So, if the column vector K contains the capital value for each bank, we can obtain the net losses as follows:

$$NL_{(sxi)} = L_{(sxi)} - (U_{sx1} \times K_{1xi})$$

The default matrix, where a binary variable is set to 1 in case of default and 0 otherwise, can be obtained according to the following formula:

$$D_{s,i}(NL_{s,i}) = \left\{ \begin{array}{ll} 1 & \text{if} \quad NL_{s,i} \geq 0 \\ 1 & \text{if} \quad NL_{s,i} < 0 \end{array} \right\}$$

and

$$D_{(sxi)} = \left\{ D_{s,i}(NL_{s,i}) \right\}$$

The excess losses matrix is then the result of the entrywise product between the NL and D matrices:

$$XL_{(s \times i)} = NL_{(s \times i)}{}^{*}D_{(s \times i)}$$

The following example (obtained following the Drehmann–Tarashev model) shows how to practically deal with this method.

The first step is the random number generation, according to a specific probability distribution. In this example, we started from a sample of standard normally distributed values, one for each considered bank plus one for the common factor (Table 2.2).

The second step is combining common and idiosyncratic factors. In this example (Table 2.3), the correlation between the two, rho, is set to 0.5 for all banks.

The resulting values are then multiplied by different values for simulating different variance levels.

The simulated losses values (Table 2.4) are positive when credit results are lower than expected (losses) and negative when results are higher than expected (negative losses = profit).

Subtracting the capital value from the initial losses, we have the net losses. Here also positive values signal actual excess losses, which means bank default if there is no recapitalization, while negative values signal a nondefault state (Table 2.5).

So, if no recapitalization is implemented, for each iteration we will have the defaults shown in Table 2.6.

From this table, we can derive important information on the bank sample. The last line reports the total number of defaults for each bank. We can see that banks 3 and 4 are the riskier, defaulting four times out of 10 simulations, while bank 5 is found to be the safest, with only one default.

On the other hand, the rightmost column reports the number of defaults for each simulation. From this vector, we can see that, hopefully, the highest frequency is for the "no default" case, but then we have two realizations for one, two, and four defaults. So, the number of defaults per iteration can be used to estimate a probability distribution, just by counting the frequency for each possible number of total defaults.

Table 2.2 Random normally distributed, uncorrelated values.

	Idiosyncratic factor 1	Idiosyncratic factor 2	Idiosyncratic factor 3	Idiosyncratic factor 4	Idiosyncratic factor 5	Common factor
Iteration 1	0.662	−0.488	−0.311	−1.376	−1.694	0.142
Iteration 2	0.038	−0.554	1.345	0.860	−0.126	0.160
Iteration 3	−0.410	−0.732	−0.357	0.267	−0.861	0.001
Iteration 4	−1.913	−0.501	0.839	−0.925	0.332	0.480
Iteration 5	−1.098	0.541	0.254	0.855	−2.097	−1.638
Iteration 6	−1.364	0.077	−0.773	0.146	−0.857	−0.348
Iteration 7	0.229	0.305	0.980	0.987	−1.474	1.202
Iteration 8	−0.095	−0.261	−1.572	−0.650	−0.124	−1.585
Iteration 9	−1.590	0.660	−0.674	1.405	−1.879	−1.004
Iteration 10	0.516	−0.222	0.782	−0.206	−1.812	2.596

Table 2.3 Random, normally distributed values, correlated with the common factor.

	Composite factor 1	Composite factor 2	Composite factor 3	Composite factor 4	Composite factor 5
Iteration 1	0.644	−0.351	−0.198	−1.121	−1.395
Iteration 2	0.113	−0.400	1.244	0.825	−0.030
Iteration 3	−0.355	−0.634	−0.309	0.232	−0.745
Iteration 4	−1.416	−0.194	0.967	−0.561	0.528
Iteration 5	−1.770	−0.350	−0.599	−0.078	−2.635
Iteration 6	−1.355	−0.107	−0.844	−0.048	−0.916
Iteration 7	0.799	0.865	1.450	1.456	−0.676
Iteration 8	−0.875	−1.019	−2.154	−1.356	−0.900
Iteration 9	−1.879	0.069	−1.086	0.715	−2.129
Iteration 10	1.745	1.106	1.976	1.120	−0.271

Table 2.4 Losses simulation.

	Bank 1	Bank 2	Bank 3	Bank 4	Bank 5	Total
Iteration 1	128.89	−35.12	−49.58	−22.41	−111.64	−89.87
Iteration 2	22.52	−39.99	311.07	16.49	−2.38	307.71
Iteration 3	−70.96	−63.36	−77.20	4.63	−59.63	−266.51
Iteration 4	−283.25	−19.38	241.67	−11.22	42.24	−29.94
Iteration 5	−353.93	−35.00	−149.72	−1.56	−210.79	−751.00
Iteration 6	−271.01	−10.71	−210.88	−0.96	−73.29	−566.85
Iteration 7	159.89	86.52	362.43	29.11	−54.08	583.86
Iteration 8	−174.99	−101.90	−538.58	−27.11	−71.98	−914.57
Iteration 9	−375.71	6.92	−271.42	14.30	−170.35	−796.27
Iteration 10	349.03	110.59	493.88	22.40	−21.70	954.20
Total	−869.53	−101.42	111.65	23.66	−733.60	

Table 2.5 Net losses.

	Bank 1	Bank 2	Bank 3	Bank 4	Bank 5	Total
Iteration 1	28.89	−80.12	−169.58	−34.41	−153.64	−408.87
Iteration 2	−77.48	−84.99	191.07	4.49	−44.38	−11.29
Iteration 3	−170.96	−108.36	−197.20	−7.37	−101.63	−585.51
Iteration 4	−383.25	−64.38	121.67	−23.22	0.24	−348.94
Iteration 5	−453.93	−80.00	−269.72	−13.56	−252.79	−1070.00
Iteration 6	−371.01	−55.71	−330.88	−12.96	−115.29	−885.85
Iteration 7	59.89	41.52	242.43	17.11	−96.08	264.86
Iteration 8	−274.99	−146.90	−658.58	−39.11	−113.98	−1233.57
Iteration 9	−475.71	−38.08	−391.42	2.30	−212.35	−1115.27
Iteration 10	249.03	65.59	373.88	10.40	−63.70	635.20
Total	−1869.53	−551.42	−1088.35	−96.34	−1153.60	

When the L matrix is filled with the excess losses (Table 2.7), obtained as the entrywise product between the defaults matrix (Table 2.6) and the net losses matrix (Table 2.5), it is possible to compute the probability distribution of losses for each bank and for the system as a whole. Since, unlike in the previous case, the losses are measured by a continuous variable, if we want to represent it by a histogram, we have to group this variable into interval values. Considering the column vector containing the crises' total losses, which is a proxy of the expected losses probability distribution, we have four cases of 0 losses, two cases of losses under 100, two cases of losses between 100 and 200, and so on.

If, instead, L is filled with the banks' covered deposits value, for the defaulted banks, summing on rows we have a set of values, which approximates the expected Deposits Insurance Schemes intervention probability distribution. With this framework, we can therefore evaluate how many crises it is possible to cover with a given DIS fund, or the fund needed for covering a given share of the expected losses. Further references on this topic are provided in Section 4.4.

Table 2.6 Resulting defaults.

	Bank 1	Bank 2	Bank 3	Bank 4	Bank 5	Total
Iteration 1	1					1
Iteration 2			1	1		2
Iteration 3						0
Iteration 4			1		1	2
Iteration 5						0
Iteration 6						0
Iteration 7	1	1	1	1		4
Iteration 8						0
Iteration 9				1		1
Iteration 10	1	1	1	1		4
Total	3	2	4	4	1	

Table 2.7 Excess losses.

	Bank 1	Bank 2	Bank 3	Bank 4	Bank 5	Total
Iteration 1	28.89					28.89
Iteration 2			191.07	4.49		195.56
Iteration 3						0
Iteration 4			121.67		0.24	121.90
Iteration 5						0
Iteration 6						0
Iteration 7	59.89	41.52	242.43	17.11		360.95
Iteration 8						0
Iteration 9				2.30		2.30
Iteration 10	249.03	65.59	373.88	10.40		698.89
Total	337.80	107.11	929.03	34.30	0.24	

2.5 Estimation of Losses Variability and Assets Riskiness

As considered when describing the Monte Carlo method, the first step for an effective simulation consists in correctly representing the input variables probability distributions. Thus, when dealing with a banking simulation, the starting point is estimating the losses variability, and the losses variability is directly linked to the assets riskiness.

If we start by considering the distribution of a panel of assets, each of which has the same probability to default, same amount, and no correlation, since the probability of default is hopefully small and the number of positions is large, the losses distribution can be represented by means of a Poisson distribution, as shown in Figure 2.6.

In a Poisson distribution, the probability of occurrence of k events (number of asset defaults, Def), with an expected rate of defaults ED, each one with probability PD, in a time interval (1 year) is given by the following formula:

$$\Pr(\mathrm{Def} = k) = \frac{\mathrm{ED}^k e^{-\mathrm{ED}}}{k!} \tag{2.4}$$

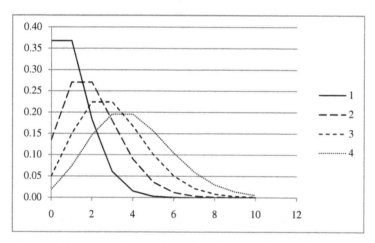

Figure 2.6 Poisson distributions for different values of ED.

Here ED represents both the expected value and the variance. So, it is the sole parameter of the distribution, and its central role for representing the process is evident.

When considering a bank assets portfolio, the problem is more complex, due to the correlation between assets, and different PD and amount of each asset. So, we have to deal with more structured distributions. Nevertheless, the assets' probability to default is fundamental, as it has a crucial role in determining both the expected value and the variance of losses, and some source is needed for determining or, at least, approximating the assets riskiness.

The evaluation of the assets' probability to default is in fact the core activity of each bank. As already considered, choosing whether to lend money to one firm or to another is based on the credit worthiness of the firms, which is to say on the estimation of its PD. The credit conditions are also directly affected by this variable, as the expected losses are to be covered by an interest spread, so the higher the asset PD, the higher the interest rate for the firm. This interest differential is used to store a reserve fund for covering the expected losses (loan losses provisions).

Thus, attention is focused on estimating the counterpart PD within banks, and the internal bank equilibriums are based on it.

Instead, not only do the publicly available data not include the PD value for each asset or for assets classes, but neither an approximation nor a weighted average of the assets PD must be included in the balance sheet. As banks are not required to declare their assets' riskiness, this value has to be estimated indirectly, based on the available variables.

Different approaches have been developed for this estimation, based on historical values of assets category riskiness, market values, capital requirements, and ratings. Another possibility is to use the loan losses provisions, as in the CAMELS and Z-score approach.

2.5.1 Sector-Historical Approach

One possibility, when data are available, is to base the estimation of loan riskiness on the activity sector and historical data on the sector performances.

This approach gives a first approximation of the bank portfolio riskiness by estimating the counterparty sector distribution and attributing to each sector the average default rate and variance registered in the previous years.

This approach, introduced by Elsinger *et al.* (2006a), has some interesting points and evident limits.

The distinction of business sectors, and the subsequent representation of each bank portfolio as a composition of loans to these sectors, brings important information on the specialization (concentration) or diversification of banks' assets, and the mixing of these sectors as risk sources gives an interesting representation of banks' business models.

Apart from the availability of the source data, which are not normally published for the large majority of countries, the major limitation in this approach is that it is backward looking, so it carries important information from the past: no references are considered with respect to business model changes, both for internal choices (change in strategy) and for the system framework changes (regulation or macrovariations).

This approach can be improved by considering business cycles, in order to correct the estimation of the assets riskiness depending on times of booms or crises, and the subsequent cyclical variation in the expected default rates. Another possible improvement can be in the evaluation of the banks' risk-taking attitude, differentiating between banks that are, for business choice, risk-takers (e.g. venture capital), and banks that are risk averse. Some reference values for these corrections can be derived from the net interest income (interests income − passive interest), from the differential level of loan losses with respect to the model-estimated values, and from the loan losses provisions levels.

2.5.2 Market Values-Based Approach

Another possibility for estimating the assets riskiness is based on market values. The hypothesis subjacent in this approach is that market values reflect the bank assets portfolio value and its expected income or losses. So, it is possible to estimate the assets' expected value and variability. As the default is due to the losses overcoming the capital coverage, and the capital level is known, simulating the assets' value variations with the correct expected value and variance, it is possible to infer the bank default risk.

This approach, proposed by Elsinger *et al.* (2006b), with information based on market values, includes both the pros and cons of its source.

In technical terms, Elsinger *et al.* (2006b) based their simulation on listed banks' market values, and then inverting the European call option pricing formula (with maturity fixed to 1 year) for deriving the banks' assets riskiness. They then simulated the system performances assuming that the banks' asset portfolio returns are normally distributed.

While the actual market value of assets is not directly observable, the market value of equity and the face value of debt for each publicly traded bank is observable. By viewing the bank equity as a European call option on the bank assets with a strike price equal to the value of debt at maturity, it is possible to derive the market value of assets for each publicly traded bank.

Starting from the Black Scholes formula for call options in one period horizon:

$$
\begin{aligned}
C(S, t) = S_t N & \left(\frac{\ln (S_t/K) + \left(r + \sigma_i^2/2 \right) T_1}{\sigma_i \sqrt{T_1}} \right) \\
& - K(t) \, e^{-rT_1} N \left(\frac{\ln (S_t/K) + \left(r + \sigma_i^2/2 \right) T_1}{\sigma_i \sqrt{T_1}} - \sigma_i \sqrt{T_1} \right)
\end{aligned}
$$

$$(2.5)$$

where $N(\cdot)$ is the cumulative standard normal distribution function. Replacing:

- the call option price $C(S, t)$ with $E_i(t)$, the equity of bank i at time t;
- the strike price K with $D_i(t) \, e^{rT_1}$, the total face value of its interest-bearing debt at time T_1,
- the spot price of the underlying asset S_t with $V_i(t)$, the value of the bank's total assets;

we have

$$E_i(t) = V_i(t)N(k_i(t)) - D_i(t)\, N\left(k_i(t) - \sigma_i\sqrt{T_1}\right) \qquad (2.6)$$

where

$$k_i(t) = \frac{\ln\left(V_i(t)/D_i(t)\right) + \left(\sigma_i^2/2\right)T_1}{\sigma_i\sqrt{T_1}} \qquad (2.7)$$

Given that $E_i(t)$, $D_i(t)$, σ_i, and T_1 are all strictly positive, the formula is invertible, and the value of total assets $V_i(t)$ can be uniquely determined. Hence, given an estimate of σ_i, it is possible to infer the market value of total assets from observable data.

As the total assets value is governed by the function,

$$V_i(T) = V_i(t-1) \times \exp\left(\left[\mu_i - \frac{1}{2}\sigma_i^2\right]h_t + \sigma_i^2(B_i(t) - B_i(t-1))\right) \qquad (2.8)$$

where $B_i(t)$ is a one-dimensional Brownian motion, and $\mathrm{Corr}\left(B_i(t), B_j(t)\right) = \rho_{ij}$, the parameters μ_i and σ_i can then be estimated simultaneously by maximizing the log-likelihood function:

$$
\begin{aligned}
L(E) = {}& -\frac{(m-1)n}{2}\ln\left(2\pi\right) - \frac{m-1}{2}\ln|\Sigma| \\
& - \sum_{t=2}^{m}\left\{\frac{n}{2}\ln\left(h_t\right) + \frac{1}{2h_t}\left(\hat{x}_t - h_t\alpha\right)' \Sigma^{-1}\left(\hat{x}_t - h_t\alpha\right)\right\} \\
& - \sum_{t=2}^{m}\sum_{i=1}^{n}\left[\ln \hat{V}_{i,t}(\sigma_i) + \ln N\left(\hat{k}_{i,t}\right)\right]
\end{aligned}
\qquad (2.9)
$$

where

$V_i(T) = V_i(0) \times \exp\left(\left[\mu_i - \frac{1}{2}\sigma_i^2\right]T + Z_i\right)$ and $\hat{V}_{i,t}$ is its estimated value;

m is the number of observations;

n is the number of banks;

$x_{i,t} = \ln\left(\frac{V_i(T)}{V_i(T-1)}\right)$, $\hat{x}_{i,t}$ is its estimated value, and $x_t = \begin{bmatrix} x_{1,t} \\ \vdots \\ x_{n,t} \end{bmatrix}$;

h_t is the time increment from $t-1$ to t;

$\alpha_i = \mu_i - \frac{1}{2}\sigma_i^2$;

$Z_i = \sigma_i B_i(T)$ is normally distributed with $E[Z_i] = 0$, $\text{Var}[Z_i] = T\sigma_i^2$ and $\text{Cov}\left(Z_i, Z_j\right) = T\sigma_i\sigma_j\rho_{ij}$

In a similar approach, Huang *et al.* (2009) estimate the (risk-neutral) probability of default (PD) of individual banks, and the asset return correlations among banks are estimated from credit default swap (CDS).

In this paper, the authors first estimate two major components that determine the risk profile of a portfolio: the probability of default and the asset return correlation. Then, they examine the dynamic linkages between default risk factors and a number of macrofinancial factors.

The estimation of the bank PD based on the CDS spreads values $s_{i,t}$ is given by

$$\text{PD}_{i,t} = \frac{a_t s_{i,t}}{a_t \text{LGD}_{i,t} + b_t s_{i,t}} \tag{2.10}$$

where $a_t = \int\limits_t^{t+T} e^{-r\tau}d\tau$ and $b_t = \int\limits_t^{t+T} \tau e^{-r\tau}d\tau$, and r is the risk-free rate.

This PD value derived from CDS values is evidently a forward-looking measure, as it reflects the average risk-neutral PD of the underlying entity during the contract.

This approach pros and cons mainly have the characteristics of the information source. On the one hand, market values are forward looking, since they are based on expectations, so they include the expected effects of all the available information. On the other hand, market values are highly variable, and therefore some smoothing is needed. Furthermore, the back-testing of market-based measures capabilities in forecasting financial crises reported its ineffectiveness, as proved by Zhang *et al.* (2015).

Moreover, the approach can only be applied to banks whose shares are traded in stock exchanges, so it can typically not be applied to small banks. This limit is more evident when considering that simulations are always system dependent, and that the small banks' balance sheets, equilibriums, and business models are typically different from those of the larger banks, so the simulation of a system that considers only the large banks can provide biased results.

2.5.3 Capital Requirements-Based Approach

In a paper aimed at DGS dimensioning, De Lisa *et al.* (2011) introduced an approach similar to that of Elsinger *et al.* (2006b), but inverting the Basel II FIRB formula (fixing maturity, loss given default, and size of the counterparty) for deriving the assets riskiness value. This also ensures that the measure is only based on balance sheet data, and it can be applied to all banks in the system, not only on the listed ones. The model (also known as SYMBOL, the acronym for systemic model for banking originated losses) was used widely for assessing the impact of European banking regulation.

In this approach, the estimation of the average default probability for the assets of any individual bank, by means of the Basel FIRB function, is based on the following reasoning.

In the Basel II framework, the capital requirement K_i of the bank i is given as the sum of the capital requirements dues for each loan of amount A_{in} to the firm n. This value is based on the probability of defaulting for the considered loan, PD_{in}:

$$K_i = \sum_n C_{in}(PD_{in}) \times A_{in} \qquad (2.11)$$

Therefore, for each bank i, there exists a value for PD, $\overline{APD_i}$, such that

$$K_i = \sum_n C_i\left(\overline{APD_i}\right) \times A_{in} \qquad (2.12)$$

Thus,

$$K_i = C_i(\overline{APD_i}) \times \sum_n A_{in} \tag{2.13}$$

and

$$\frac{K_i}{\sum_n A_{in}} = C_i(\overline{APD_i}) \tag{2.14}$$

where K_i is the capital requirement for bank i and $\sum_n A_{in}$ is the same bank's total assets.

The average asset probability of defaulting for bank i ($\overline{APD_i}$) is computed as the PD that allows the actual value of the capital requirement for that specific bank, K_i (extracted from balance sheet data), to be equal to its numerically calculated value:

$$\overline{APD_i} : \left[C_i(\overline{APD_i}) = \frac{K_i}{\sum_n A_{in}}\right] \tag{2.15}$$

$C_{in}(PD_{in})$ is obtained as the result of a capital allocation formula.

The Basel FIRB formula sets a capital coverage of 99.9% of the undiversified component, formalized by the following formula, which includes the loan loss given default LGD_{in}, the residual time to maturity M_{in}, and the firm size S_n:

$$C_{in}(PD_{in}, LGD_{in}, M_{in}, S_n) =$$

$$\left[LGD_{in} \times N\left[\sqrt{\frac{1}{1 - R(PD_{in}; S_n)}}N^{-1}(PD_{in})\right.\right.$$

$$\left.\left.+ \sqrt{\frac{R(PD_{in}; S_n)}{1 - R(PD_{in}; S_n)}}N^{-1}(0.999)\right] - PD_{in} \times LGD_{in}\right]$$

$$\times \frac{1 + (M_{in} - 2.5) \times B(PD_{in})}{1 - 1.5 \times B(PD_{in})} \times 1.06 \tag{2.16}$$

where

$$B(PD_{in}) = [0.11852 - 0.05478 \ln (PD_{in})]^2 \tag{2.17}$$

and

$$R(PD_{in}; S_n) = 0.12 \frac{1 - e^{-50 \times PD_{in}}}{1 - e^{-50}}$$

$$+ 0.24 \left[1 - \frac{1 - e^{-50 \times PD_{in}}}{1 - e^{-50}} \right] - 0.04 \left[\frac{S_n - 5}{45} \right]$$

$$(2.18)$$

thus depending even on the loan loss given default (LGD), maturity (M), and firm size (S).

Setting these variables to their standard values, that is, loss given default (LGD $= 0.45$), maturity ($M = 2.5$) and firm size ($S = 50$), it is possible to invert the formula for a proxy of the average assets PD of the bank APD:

$$\overline{APD_i} : C_i \left(\overline{APD_i} | LGD = 0.45; M = 2.5\ S = 50 \right) = \frac{K_i}{\sum_n A_{in}}$$

$$(2.19)$$

The obtained $\overline{APD_i}$ values are then used to generate a set of correlated losses across all banks in the system.

As the FIRB formula is based on a VaR, it is implicitly based on a probability distribution for losses. The capital covers the losses up to the VaR threshold, set in this case at 99.9%.

For each simulation s, bank i's losses L_{is} are computed using a Monte Carlo simulation based on the FIRB formula, modified for having the random variable z_{is} replacing the VaR (capital coverage) threshold:

$$L_{is}\left(z_{is};\ \overline{APD_i}\right) = \left[0.45 \times N \left[\sqrt{\frac{1}{1 - R\left(\overline{APD_i}; 50\right)}} N^{-1}\left(\overline{APD_i}\right) \right. \right.$$

$$\left. + \sqrt{\frac{R\left(\overline{APD_i}; 50\right)}{1 - R\left(\overline{APD_i}; 50\right)}} N^{-1}(z_{is}) \right] - \overline{APD_i} \times 0.45 \right]$$

$$\times \frac{1}{\left(1 - 1.5 \times B\left(\overline{APD_i}\right)\right)} \times 1.06$$

$$(2.20)$$

Here $N^{-1}(z_{is}) \sim N(0, 1)$ $\forall i, s$ is the random variable representing the nondiversifiable shock that is correlated over banks so that $\text{cov}(z_{is}, z_{ls}) = 0.5$ $\forall i \neq l$.

Here also the approach mainly has the characteristics of its information source. The risk weighting made by bank is in this case the main source of information. This evaluation is made by means of an internal methodology, which can only be monitored by the supervisors, by comparing the actual losses with the expected losses. As this monitoring is not on the direct measure of risk, but on the indirect effects of risk on losses, the main forward-looking evaluation for supervisors is in its methodology assessment.

This ensures good homogeneity when dealing with a system supervised by the same authority, but there are several obvious concerns such as the actual homogeneity of assessment made by different supervisors. So, when dealing with international panels (e.g., the eurozone), more caution is advised.

There are additional doubts about the balance sheet policy, since around the time of an actual crisis, managers sometimes try to do some "window dressing", so as to hide the actual riskiness of some (actually defaulted or near to default) loans. This and other actions aimed at underestimating risks can bias the proper measurement of the actual riskiness and evidently lead to biased simulations.

2.5.4 Ratings-Based Approach

Drehmann and Tarashev (2013) evaluated the systemic risk contribution of banks and developed an approach similar to that of Elsinger *et al.* (2006b), with Monte Carlo simulated common shocks, based on setting the idiosyncratic and common shocks variance (and covariance matrix) to obtain an a posteriori default rate coherent with the estimates of the banks' probability of default as published by rating agencies.

The DT model simulates losses according to the following formula:

$$L_{is}(m_s, z_{is}, \rho_i) = \rho_i m_s + \sqrt{1 - \rho_i^2} z_{is} \tag{2.21}$$

where $\rho_i \in [0, 1]$ is the common factor loading, i.e. the correlation between the common factor and the simulated losses, and m_s and z_{is} are the common and idiosyncratic shocks, mutually independent normal variables with zero means and equal variances.

In the cited paper, both the common and idiosyncratic exogenous shocks are generated with a variance calibration so that the resulting bank PD exactly matches the corresponding 1-year expected default frequency (EDF), estimated by Moody's KMV for 2006–2009. Note that if the probability of default estimated by the rating agencies also includes contagion risks, as contagion risks are system dependent, this way of setting the losses variance becomes system dependent. So, in this case, the approach can be directly applied only when simulations refer to the same system as considered by the rating agencies for estimating its contagion risks. For different (hypothetical) systems, or when including banks with no rating, the parameters are to be calibrated on the basis of some adaptation of this rule.

The limits of this approach are similar to those of market-based measures, in terms of forecasting capabilities (as proved by the ratings of some defaulted banks, for example, Lehman Brothers, just before the actual default), and for the extension to small banks.

2.5.5 CAMELS—Z-Score Approach

The possibility of deriving a bank PD value from balance sheet data, so as to apply it to all commercial banks, has already been explored, in particular by Altman since 1977 (see Altman, 1977; Altman and Saunders, 1998; Altman *et al.*, 2014) with the so-called Z-score model and its subsequent evolutions.

The Z-score approach was developed as a multivariate credit scoring system, based on a linear function of accounting and market variables that best distinguishes between defaulting and nondefaulting firms. The parameters estimation for maximizing the predictive power of the variables is based on a discriminant analysis. This methodology yielded very interesting results, even if it needs specific tuning for each sector, time, and economic framework.

The discriminant models had two important applications in banking. The first one is for a computer-based creditworthiness assessment, which has had a widespread use in consumer credit, but also as a support for firms' credit assessment.

The second application is in predicting bank failures. With this aim, the approach was found to be similar to the CAMEL approach, a supervisory rating system developed by the U.S. bank regulators in the early 1980s.

CAMEL(S) is the best-known example of risk-rating models based on accounting variables. Its name is an acronym that relates to the dimensions of bank conditions assessed by the system, namely, capital adequacy, asset quality, management quality, earnings, liquidity, and (from 1996) sensitivity to market risk. Many studies tested the accounting variables proxying the CAMELS components performances as predictors of bank distress (see Sahajwala and Van den Bergh, 2000; Cole and White, 2012). The variables typically tested for assessing the bank probability to default (expressed in terms of its incidence on total assets, except from log of bank total assets) are total equity, loan loss reserves, return on assets, nonperforming assets, securities held for investment plus securities held for sale, brokered deposit, log of bank total assets, cash and items due from other banks, intangible assets, real estate mortgages, commercial and industrial loans, and consumer loans.

The probability of default of each can be estimated by means of a probabilistic model, based on each bank-specific characteristic, $x_{i,k}$:

$$PD_i = \sum_k \beta_k x_{i,k} + \epsilon_i \tag{2.22}$$

In the literature, the highest discriminative power is often found in asset quality, capital adequacy, and management quality.

By estimating the probability of the bank defaulting, and considering the value of its capitalization, it is possible to derive the bank's assets riskiness as follows.

Starting from the estimated bank PD, BPD_i, and the bank actual capital K_i, we have

$$K_i = \sum_n C_{in}(PD_{in}, (1 - BPD_i)) \times A_{in} \tag{2.23}$$

Thus, there exists a value for PD, $\overline{APD_i}$, so that

$$\overline{APD_i} : C_i(\overline{APD_i}, (1 - BPD_i)) = \frac{K_i}{\sum_n A_{in}} \tag{2.24}$$

where $\sum_n A_{in}$ is for bank i's total assets.

In fact, the estimated $\overline{APD_i}$ can better be considered as a proxy of the bank activity riskiness, instead of the bank assets riskiness, as the bank PD obtained from ratings, market value, or z-scores is estimated on the basis of different aspects of the bank's activity, not only on the bank's assets riskiness.

2.6 Simulating Shocks: Correlated Risk Factors

The next step, after analyzing the case of the unprecedented failure of one bank, is considering the consequences of the failure of more banks.

Even if a multiple bank default can happen for the same reasons hypothesized before for the single case, it is extremely unlikely to happen without a common shock hitting the system. So the extension of this exercise, instead of just raising the number of initial failures hypothesized, is to develop a thorough analysis of how and why banking crises happen, determine how banks are (at least partially) exposed to the same shocks, and model these mechanisms.

After exploring the risks and the common exposure to some sources of risk for different banks, two approaches are used for simulating the system.

The first one is based on an exogenous shock, affecting more (possibly all) banks at the same time, and results in stress testing the system (see Section 2.2). The second is based on the Monte Carlo simulation (presented in Section 2.4.1).

The first approach, being based on a single exogenous shock, can sound as just an extension of the previous framework, but even this apparently simple difference, the shock hitting all the banks instead of only one, induces deep changes on both the representation of the simulated system and its mechanisms.

With reference to the system representation, banks are, exposed to the same risk sources. This happens since, in particular when considering a domestic banking system, the banks are acting in the same market, thus financing firms belonging to the same economic system and exposed to the same business cycle. Quite often, different banks are also financing the same firms.

The exposure to the same business cycle means that borrowers' failures are more likely to happen in some years (crises) than in others (booms), and this affects all banks' results, if not in an identical manner, at least in similar ways. But banks are also exposed to other risks, like exchange rate variations, interest rate variations, government policies, and so on. This also exposes banks to common shocks, and here also the effects are different for different banks, due to their different credit and market risk exposures.

The effect of this correlated exposure, and therefore the effect of common shocks, translates not only into the possibility that more banks default but also in the weakening of a large part of the banking system. Even if only one bank defaults as a result of a direct shock, the other banks are more likely to suffer from interbank losses since their capital (loss-absorbing capacity) is already reduced by the common shock. It is evident that the larger the number of affected banks, and the deeper the impact of the shock on affected banks, the higher the risk of contagion.

The representation of this framework can include different levels of complexity, depending on the object of the analysis.

One possible way to model this framework is to start from one random variable for each considered bank, and then mixing them up to simulate correlated variables affecting banks' results.

This approach, actually used in De Lisa *et al.* (2011), gives a correct representation of the correlated shocks, even if it does not allow us to explicitly include the common risk sources (Figure 2.7).

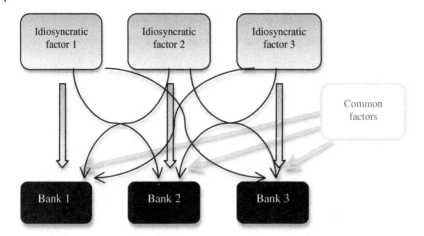

Figure 2.7 Banking system simulation: correlated idiosyncratic shocks.

This representation can be numerically simulated starting from the random generation of multiple correlated variables. If the available tools do not allow the direct generation of multiple correlated variables, it can be obtained by generating uncorrelated variables, and then multiplying them by the Cholesky decomposition of the correlation matrix.

In practice, the envisaged correlation matrix Corr:

$$\text{Corr}_{(i \times i)} = \begin{pmatrix} 1 & \rho_{1,2} & \rho_{1,3} & \cdots & \rho_{1,i} \\ \rho_{1,2} & 1 & \rho_{2,3} & \cdots & \rho_{2,i} \\ \rho_{1,3} & \rho_{2,3} & 1 & & \rho_{3,i} \\ \vdots & \vdots & & \ddots & \vdots \\ \rho_{1,i} & \rho_{2,i} & \rho_{3,i} & \cdots & 1 \end{pmatrix} \qquad (2.25)$$

is decomposed to give the lower triangular matrix Chol:

$$\text{Chol}\left(\text{Corr}_{(i \times i)} =\right) \begin{pmatrix} \text{Chol}_{1,1} & 0 & 0 & \cdots & 0 \\ \text{Chol}_{2,1} & \text{Chol}_{2,2} & 0 & \cdots & 0 \\ \text{Chol}_{3,1} & \text{Chol}_{3,2} & \text{Chol}_{3,3} & & 0 \\ \vdots & \vdots & & \ddots & \vdots \\ \text{Chol}_{i,1} & \text{Chol}_{i,2} & \text{Chol}_{i,3} & \cdots & \text{Chol}_{i,i} \end{pmatrix}$$

$$(2.26)$$

Multiplying the Chol matrix by a column vector N produces a sample vector with the proper correlation properties,

$$\text{Chol}_{(i \times i)} \cdot N_{(i \times 1)} = R_{(i \times 1)} \tag{2.27}$$

or which is the same but more coherent with the representation used,

$$N_{(1 \times i)} \cdot \left(\text{Chol}_{(i \times i)}\right)^T = R_{(1 \times i)} \tag{2.28}$$

which for a sample of s row vectors becomes

$$N_{(s \times i)} \cdot \left(\text{Chol}_{(i \times i)}\right)^T = R_{(s \times i)} \tag{2.29}$$

It is worth remembering that the Cholesky decomposition can only be applied to positive-definite matrices. In fact, the covariance matrix of a multivariate probability distribution is always positive semidefinite, thus not always suitable for the Cholesky decomposition. A simple test for verifying if the correlation matrix is positive definite is to compute its eigenvalues: If all its eigenvalues are positive, the matrix is positive definite.

When considering a correlation matrix estimated from actual data, it can happen that some of its eigenvalues are zero. This typically happens when the number of observations the correlation matrix is derived from is lower than the number of variables. In this case, a solution of the problem can be oversampling.

When, instead, the correlation matrix is obtained from a theoretical model (see Section 2.8), it is worth testing it for the actual properties of the matrix and also its suitability for this use. Anyway, a correlation matrix with 1 as diagonal elements and a homogeneous positive value for all other correlations among banks is always positive definite.

So, we will start with the correlation matrix, as provided in Table 2.8.

The resulting Cholesky decomposition will be as provided in Table 2.9.

Multiplying the idiosyncratic factors reported in Table 2.2 by the decomposition matrix, we will have the correlated idiosyncratic factors, as provided in Table 2.10.

Table 2.8 Uniform correlation matrix.

	Factor 1	Factor 2	Factor 3	Factor 4	Factor 5
Factor 1	1	0.5	0.5	0.5	0.5
Factor 2	0.5	1	0.5	0.5	0.5
Factor 3	0.5	0.5	1	0.5	0.5
Factor 4	0.5	0.5	0.5	1	0.5
Factor 5	0.5	0.5	0.5	0.5	1

Table 2.9 Cholesky decomposition of the uniform correlation matrix.

1	0	0	0	0
0.5	0.8660	0	0	0
0.5	0.2887	0.8165	0	0
0.5	0.2887	0.2041	0.7906	0
0.5	0.2887	0.2041	0.1581	0.7746

Table 2.10 Correlated idiosyncratic factors.

	Factor 1	Factor 2	Factor 3	Factor 4	Factor 5
Iteration 1	0.662	−0.091	−0.064	−0.961	−1.403
Iteration 2	0.038	−0.461	0.957	0.813	0.172
Iteration 3	−0.410	−0.839	−0.708	−0.278	−1.114
Iteration 4	−1.913	−1.390	−0.416	−1.661	−0.819
Iteration 5	−1.098	−0.080	−0.185	0.335	−1.830
Iteration 6	−1.364	−0.615	−1.291	−0.702	−1.458
Iteration 7	0.229	0.379	1.003	1.183	−0.583
Iteration 8	−0.095	−0.274	−1.407	−0.958	−0.642
Iteration 9	−1.590	−0.224	−1.155	0.369	−1.975
Iteration 10	0.516	0.066	0.833	0.191	−1.082

Where each column is correlated with each of the others, the higher the number of lines (iterations), the smaller the difference between the theoretically induced correlation and the actually realized one.

The actual simulation of losses must then include the corrections for considering the bank size and assets riskiness.

2.7 Simulating Shocks: Combining Idiosyncratic and Common Shocks

Finally, a complete model should include each bank risk as determined by two components: the first due to its specific activity (idiosyncratic component) and the second due to the exposure to the general economic business cycle (common factor) (Figure 2.8).

This representation was explored in a number of papers, such as Drehmann and Tarashev (2013). In their paper, the losses are simulated by means of the following formula:

$$L_{is}(m_s, z_{is}, \rho_i) = \rho_i m_s + \sqrt{1 - \rho_i^2} z_{is} \qquad (2.30)$$

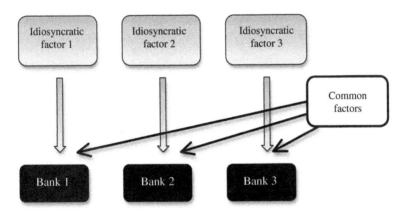

Figure 2.8 Banking system simulation: idiosyncratic and common shocks.

Table 2.11 Correlation matrix between common factor and idiosyncratic factor.

	Common factor	Idiosyncratic factor
Common factor	1	ρ_i
Idiosyncratic factor	ρ_i	1

where $\rho_i \in [0,1]$ is the common factor loading, m_s is the randomly generated value for the common factor, and z_{is} is the randomly generated idiosyncratic factor for each of the i banks.

The formula above exactly corresponds to computing correlated values from the idiosyncratic and common factors (Table 2.11), by means of the correlation matrix:

Cholesky decomposition gives the matrix in Table 2.12.

This model also allows one to consider, for each bank, a different value of correlation between the idiosyncratic and common factors, without the complexity (such as the non-positive definiteness) related to the estimation of only one matrix that includes all the correlations between the considered banks. It is also worth noting that, while the initial random variables are uncorrelated, since the resulting factors are correlated with the common factor, they also become correlated among themselves:

If

$$m, z_1, z_2 \sim N(0, 1)$$

Table 2.12 Cholesky decomposition of the correlation matrix between a common factor and an idiosyncratic factor.

	Common factor	Idiosyncratic factor
Common factor	1	0
Idiosyncratic factor	ρ_i	$\sqrt{1 - \rho_i^2}$

and

$$\text{Corr}(z_1, z_2) = 0$$
$$\text{Corr}(z_1, m) = 0$$
$$\text{Corr}(m, z_2) = 0$$
$$L_1 = \rho_1 m + \sqrt{1 - \rho_1^2} z_1$$
$$L_2 = \rho_2 m + \sqrt{1 - \rho_2^2} z_2$$

then

$$\text{Corr}(L_1, L_2) = \rho_1 \rho_2$$

Co-Pierre (2013) compared systemic risk caused by contagion with the risk triggered by common shocks. In the case of pure interbank contagion, the largest bank in the system is selected and exogenously sent into default (caused, for example, by an idiosyncratic shock to the banking capital). In the event of a common shock, all banks suffer a simultaneous loss of a fraction of their banking capital. When a common shock hits the system, it causes banks with insufficient capitalization to default. Even if only a small number of banks default, a larger number of banks become more vulnerable, and this can lead to a large number of defaults. Thus, the impact of a common shock on the system is more severe than in the contagion case.

2.8 Correlation

Another fundamental issue, when dealing with common shocks, refers to correlation. In fact, correlation and contagion are closely connected. A simultaneous shock not only affects the defaulting bank(s) but also weakens their creditor banks, so contagion is more likely to start. Following Elsinger *et al.* (2006), "Among the two driving sources of systemic risk, the correlation in exposures is far more important than financial linkages." A test in De Lisa *et al.* (2011) confirms these findings.

Figure 2.9 Multilayer correlation structure.

The correlation structure, in fact, includes at least three different levels: the correlation between assets in one bank portfolio, the correlation between results of banks operating in the same country, and the correlation between macrovariables of countries (or states) in the same economic system (Figure 2.9).

The first level refers to the correlation among loans issued by the same bank (assets—bank level).

In fact, since banks normally operate in specific areas and targets, the borrowers of the bank loans are exposed to the same economic framework, and their results are at least partially due to the economic business cycle, so typically some years register more loan defaults than others. This correlation mostly affects banks that operate in a single business sector, so the banks can reduce the exposure to this risk by diversifying their loans to different sectors. In any case, even the highest diversification cannot fully cut out the correlation (Figure 2.10).

Within the Basel II framework, the FIRB formula includes some important references to the determinants of the correlation between assets.

The first important point is the recognition of the customer size as a relevant factor with respect to correlation.

In this model, based on statistical analysis, the higher dimension of the loan counterpart was acknowledged as determining a higher correlation. In fact, the results of larger firms are

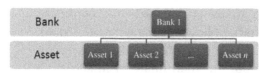

Figure 2.10 Correlation structure: single bank level.

statistically more stable than the results of smaller firms, and this is possibly because the larger firms often include diverse branches or diverse product lines. Thus, the resulting returns include some pooling, partially compensating among them, and thereby smoothing the specific effects of each single business line and enhancing the components more correlated to the general business cycle.

A similar effect is considered in the same FIRB modeling with reference to customer riskiness. For exposures characterized by high default probability, defaults tend to be driven more by specific (diversifiable) factors than by systemic factors, registering a lower correlation than for lower PD values, where exposures tend instead to be more dependent on the business cycle.

Even if this analysis and modeling were important to correctly evaluate the capital that a bank needs in order to cover the risks, when dealing with a simulation of banks or banking systems this layer is typically not considered. This is because the correlation between single loans cannot be observed or estimated, given that this type of information is not available outside the bank. Nevertheless, this analysis provides some clues about how to model the further layers of correlation.

What is relevant is that the correlation among exposures induces a correlation between bank results and the general business cycle.

The linkage to macroeconomic factors implies a second layer of correlation (banks—country level) (Figure 2.11).

With reference to the literature, already in Elsinger *et al.* (2006a) for Austria, correlation in exposures is acknowledged to possibly induce contemporaneous crises in different banks, with important systemic effects.

In their work, a positive correlation is included in the model by drawing the same quantile of the average default frequency distribution of each bank. The correlation coefficients of the

Figure 2.11 Correlation structure: domestic system level.

individual losses from credit risk to the aggregate loan losses turn out to have a mean value of 0.4.

A more flexible way of modeling correlation among banks can be simply including two factors, a common one and an idiosyncratic one, with adequate weights, into the bank artificial results generation. It is clearly important to determine the appropriate weights.

The simplest way involves fixing a standard value for all banks, a solution quite often used in the literature. Often, the common factor weight is defined in order to produce a final correlation of around 0.5.

Similarly, some other authors use a direct generation of correlated shocks, with a correlation of about 0.5. In fact, the estimations in the literature quite often report estimated values around 0.5 for the correlation between bank results, based on either stock market values or balance sheets values (see Table 2.13 as an example).

A more sophisticated modeling approach that is found in other papers uses different correlation values for each bank. This is clearly a more realistic approach, requiring a more precise analysis of banks' results. The single bank correlation to the common factor can be estimated in two ways.

The first way (see Hull and White, 2004) consists in using the equity return correlation as an approximation for the asset return correlation, thus to base it on historical values, and then in assigning each bank a correlation factor equal to the actual correlation registered in the previous years. This backward-looking approach guarantees a correct approximation of the correlation factor based on what happened in the previous years.

The second way is to analyze and model the historical correlation distribution, possibly by employing the same references of the Basel II FIRB formula for loans, and therefore setting lower correlation values (lower common factor loadings) for the less correlated categories (smaller banks and higher PD) and higher weights of the common factor for the more correlated categories (larger banks and lower PD), and, when available, on the basis of the assets categorization of each bank.

Table 2.13 Correlation between ROE of a panel of German banks for 2007–2013.

	DB	CMB	DZ	BL	LBW	NLG	LHTG	DDG	HSH	LBH
DB	1.000	0.116	0.498	0.443	0.494	0.473	0.651	0.703	0.706	0.568
CMB	0.116	1.000	0.203	0.660	0.537	0.643	0.251	0.278	0.597	−0.047
DZ	0.498	0.203	1.000	0.763	0.814	0.060	0.825	0.835	0.680	0.035
BL	0.443	0.660	0.763	1.000	0.918	0.548	0.858	0.689	0.916	0.033
LBW	0.494	0.537	0.814	0.918	1.000	0.488	0.906	0.589	0.815	−0.139
NLG	0.473	0.643	0.060	0.548	0.488	1.000	0.471	0.178	0.596	−0.029
LHTG	0.651	0.251	0.825	0.858	0.906	0.471	1.000	0.675	0.827	0.033
DDG	0.703	0.278	0.835	0.689	0.589	0.178	0.675	1.000	0.785	0.492
HSH	0.706	0.597	0.680	0.916	0.815	0.596	0.827	0.785	1.000	0.392
LBH	0.568	−0.047	0.035	0.033	−0.139	−0.029	0.033	0.492	0.392	1.000

Then, one can properly compute each bank correlation to the common factor based on variable mapping or estimation based on the actual reference variables. This approach requires a more detailed analysis and modeling, but it allows the role of the fundamental variables to be estimated before applying them to the actual bank situation based on the last balance sheet values, in order to have a forward-looking estimation.

Using the macrovariables as a proxy allows the effects of the country determinants on banks and banking systems to be modeled and verified.

Some papers have analyzed how the GDP variations influence the banking results, loans riskiness, and losses. Karimzadeh *et al.* (2013), Demirguc-Kunt and Huizinga (2000), and Bikker and Hu (2002) found that bank profits are correlated with the business cycle. Athanasoglou *et al.* (2008) verified that some important variables affecting bank results are correlated with the business cycle, so it is not evident if the correlation with the business cycle is indirectly carried by these variables, without a specific consideration of these variables' effects. Even after controlling for the effect of other determinants, their findings confirm that the business cycle significantly affects bank profits. Albertazzi and Gambacorta (2009) specified that bank profits' procyclicality is due to two different channels: the first referring to the lending activity, positively affecting the net interest income, and the second due to the assets quality, negatively linked to loan losses and loan loss provisions.

The third layer of correlation refers to the international linkages (countries—system level). When dealing with systems including more than one country, it is also necessary to evaluate the correlation among the different countries (Figure 2.12).

Figure 2.12 Correlation structure: international level.

Moreover, SIFIs or other large banks typically act in different countries and have foreign interbank counterparts. Thus, these banks are influenced by international factors, and to obtain reliable results it is important to reproduce both the correct exposure of the bank to each national factor and the correlation between the considered national factors.

Even when considering the simplest correlation between countries as approximated by their correlation of the GDP variation (Table 2.14), we observe that the mappings do matter, and the simulation results are deeply affected by this specification. A test in Zedda (2015) confirms that the simulation results, depending on the simulation settings, can be up to 10 times higher than the uncorrelated results.

This mapping can also be reproduced starting with a set of uncorrelated variables, one for each considered country, and then correlating them by means of the Cholesky decomposition of the estimated correlation matrix, or starting from one idiosyncratic variable for each country plus one common factor, as for the banks–country layer.

The representation can also be simplified by a factor analysis. The results in Table 2.15 show that by considering the first three factors only, one can already explain more than 95% of the actual variability.

It is thus possible in this case to have a good approximation of the system's expected results starting from three uncorrelated variables and computing the simulated variation for each country on the basis of the coefficients estimated within the factor analysis. So, when including banks operating in different countries, the model should include the two layers of correlation.

In this way it is also possible to correctly represent the international activity of large banks acting in different countries, which are not influenced by single country determinants, their results being rather related to a pooling of the exposures to different countries.

One final interesting question concerns the diversification process: Is it safer to have a system with specialized banks, i.e. to have a clear diversification among banks, or a system where all banks have the best diversification in their assets portfolio?

Table 2.14 Correlations between GDP growths of several eurozone countries for 2004–2014 based on Eurostat data.

	Austria	Belgium	Cyprus	Estonia	Finland	France	Germany	Greece	Ireland	Italy
Austria	100%	91%	66%	86%	98%	96%	90%	39%	76%	92%
Belgium	91%	100%	71%	80%	96%	95%	83%	51%	78%	96%
Cyprus	66%	71%	100%	38%	69%	58%	43%	61%	42%	66%
Estonia	86%	80%	38%	100%	84%	88%	78%	29%	90%	80%
Finland	98%	96%	69%	84%	100%	97%	89%	45%	77%	96%
France	96%	95%	58%	88%	97%	100%	88%	38%	78%	95%
Germany	90%	83%	43%	78%	89%	88%	100%	16%	65%	88%
Greece	39%	51%	61%	29%	45%	38%	16%	100%	57%	48%
Ireland	76%	78%	42%	90%	77%	78%	65%	57%	100%	80%
Italy	92%	96%	66%	80%	96%	95%	88%	48%	80%	100%

Table 2.15 Factor analysis computed eigenvalues of the GDP correlation matrix.

Eigenvalue	%	Cumulative %
7.7289	77.29	77.29
1.1487	11.49	88.78
0.6515	6.52	95.29
0.2135	2.13	97.43
0.1127	1.13	98.55
0.0930	0.93	99.48
0.0273	0.27	99.76
0.0184	0.18	99.94
0.0052	0.05	99.99
0.0007	0.01	100.00

In fact, the second case brings to a safer stabilty of single banks, but also to a higher correlation among banks and so a higher risk of systemic crises. Some references on this topic are given in Section 4.8 (Figure 2.13).

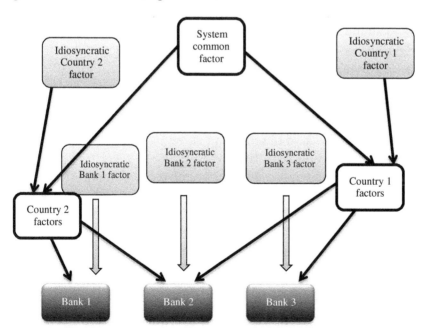

Figure 2.13 Banking system simulation: two-layer common factors.

2.9 The Interbank Matrix

The first analyses in the literature were devoted to analyzing the possible effects of the hypothetical failure of a single institution, isolating one aspect of the problem: the consequences of a failure, in particular on banks linked to the failing one. This revealed the importance of a proper understanding of the possible contagion mechanisms and of their determinants.

As specified by Degryse and Nguyen (2007), contagion results from the materialization of two risks: the risk that at least one bank is hit by a shock (likelihood of a shock) and the risk that this shock propagates through the system (potential impact of the shock).

While the first risk, the likelihood of a shock, was analyzed in the previous sections, it is now important to analyze the importance of the shock propagation and of the channels allowing it.

The first, fundamental, channel of impact is given by the direct interbank lending. If a bank defaults, the creditor banks will not have their money back, or, at least, the creditor banks cannot have it according to the terms agreed.

Interbank lending is one fundamental way for managing banks' liquidity and risks.

Banks rely on intermediaries for a variety of functions. One fundamental function is liquidity management, as the interbank lending allows channeling funds from surplus banks to deficit banks (e.g., Niehans and Hewson, 1976; Bruche and Suarez, 2010).

Other important functions of interbank exposures are for balancing funding needs and the actual capability of attracting depositors, and for the maturity profile management, lending or borrowing money at different maturities for matching their initial maturity profile with the target.

This use of the interbank market has given banks important flexibility, and thus a higher efficiency, leading the actual bank systems in fact to depend on the access to this market.

The fact that the counterparts are banks, and thus firms with a high rating, with a typically low default risk, has eased its use and sometimes led to an unjustifiably optimistic exposure to other banks.

The recent financial crisis, which actually started contagion effects among banks of the more developed countries, prompted the financial system to reopen its eyes to the fact that banks are in any case exposed to default risk; thus before acting in the interbank market it is worth analyzing the creditworthiness of the counterpart, even if it is a bank.

With reference to the representation of the effects to the creditor bank in case of interbank default, the basis for its quantification is the value of the creditor bank exposure to the defaulting one. Moreover, the actual loss share for the creditor, called the loss given default (or LGD), has to be quantified.

This raises two questions:

- Are there any available data on interbank lending for evaluating the exposure of a bank to each other?
- How is it possible to estimate the loss for a creditor bank given the default of the debtor bank?

In mathematical terms, the problem can be represented as follows:

$$L_{ij} = IB_{ji} \times LGD_{ji} \tag{2.31}$$

where

L_{ij} is the loss of bank i as a consequence of bank j's default;
IB_{ji} is the interbank credit (exposure at default) of bank i with regard to bank j;
LGD_{ji} is the loss of bank i given default of bank j.

So, the following two questions arise:

- How do we quantify IB_{ji} for all i and j?
- How do we estimate LGD_{ji}?

The analysis of the interbank matrix has revealed some important findings even with reference to its structure, besides the actual quantification of the exposure values. In particular, some papers based on the network theory approach (see Allen and Gale, 2000 and Freixas et al., 2000) have shown that the probability of contagion depends on the structure of the interbank market. Allen and Gale considered a banking system with different lending structures and showed that for the same shocks some structures would result in contagion while others would not.

These papers analyzed the interbank network structures, on the basis of the actual existence of linkages between banks, distinguishing a "complete" structure, where every bank has symmetric exposures to all other banks (corresponding to a matrix with no zero elements except for the diagonal), from an "incomplete structure," where each bank is linked only to some other bank (so that some elements in the interbank matrix are zero). They also distinguish "connected" structures, where each bank is indirectly linked to all others, from disconnected structures, where banks are grouped and linked within the group but no connection exists between groups.

Another important structure of the interbank lending is the money-centric structure, in which the peripheral banks are linked to the bank at the center but not to each other.

Freixas *et al.* (2000) analyzed how contagion can spread in a system with a money-centric structure and found that the results depend on the model parameters.

More network analysis results show that disconnected structures are more prone to contagion than "complete" structures, but they prevent contagion from spreading to all banks.

The different structures of the interbank matrix (Figure 2.14 and Tables 2.16–2.18) have an important impact on contagion. In fact, the maximum entropy ensures that the initial default effect is spread to the whole system. In a homogeneous system, where each bank has the same value of capital, assets, and interbank exposures, either no bank defaults or the whole system defaults. When banks have diversified capitalization and exposure levels, the effect is smoother, but, all the same, shocks are propagated in the entire system, and the impact of an initial default has a reduced effect, hitting all banks.

On the other hand, when the exposures are more concentrated, the banks hit by contagion take a more significant plunge, but not all banks are hit by every initial default.

The two effects, spreading a default on more banks or influencing fewer banks but more seriously, in someway compensate,

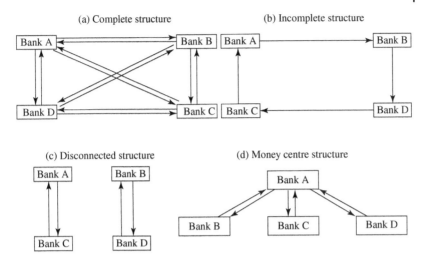

Figure 2.14 Stylized network structures of the interbank market. *Source:* Upper, 2011. Reproduced with permission of Elsevier.

as the higher impact can induce more defaults by contagion, so the process restarts possibly by involving banks that were not linked to the first defaulted bank.

Some theoretical analyses were devoted to this problem, such as in Brusco and Castiglionesi (2007) and Hasman and Samartin (2008). Their results showed that incompletely connected markets have a higher resilience to contagion effects.

Table 2.16 Complete structure.

	Bank 1	Bank 2	Bank 3	Bank 4	Bank 5		IB*d*
Bank 1	0	1	4	1	2		8
Bank 2	1	0	1	1	1		4
Bank 3	1	3	0	1	2		7
Bank 4	1	2	2	0	2		7
Bank 5	1	1	1	1	0		4
							30
IB*c*	4	7	8	4	7	30	

Table 2.17 Incomplete connected structure.

	Bank 1	Bank 2	Bank 3	Bank 4	Bank 5		IB*d*
Bank 1	0	2	3	0	3		8
Bank 2	1	0	2	0	1		4
Bank 3	0	3	0	2	2		7
Bank 4	3	0	3	0	1		7
Bank 5	0	2	0	2	0		4
							30
IB*c*	4	7	8	4	7	30	

In fact, the structure resilience cannot be considered irrespective of any reference to the other bank characteristics. The step effect that characterizes contagion is in this case crucial. The consequences of a default on a system with low interbank exposures fully diversified results in transmitting a small loss to every bank; so if the banks are well capitalized, none will default by contagion. If, instead, the interbank exposures are higher and the capitalization level is low, the initial shock will propagate through the system, leading possibly to a default of all banks.

From another point of view, Ladley (2011) argued that interbank markets are characterized by a knife-edge property, so the

Table 2.18 Incomplete disconnected structure.

	Bank 1	Bank 2	Bank 3	Bank 4	Bank 5		IB*d*
Bank 1	0	3	5	0	0		8
Bank 2	1	0	3	0	0		4
Bank 3	3	4	0	0	0		7
Bank 4	0	0	0	0	7		7
Bank 5	0	0	0	4	0		4
							30
IB*c*	4	7	8	4	7	30	

small shocks reveal a stabilizing effect of the interbank exposures. Instead, large shocks can induce contagion effects that are eased by a large interbank market volume.

The same two scenarios may have different effects in case of a higher concentration of interbank exposures. The first case can in any case induce some default by contagion, as the concentration in exposures may destabilize the counterpart bank, while in the second case the incompleteness of the market and the lack of contagion channels can induce higher resilience.

Mistrulli (2010) did an interesting test on the actual contagion conductivity of complete and incomplete markets. In his analysis, he compared the maximum entropy and one actual interbank matrix obtained from confidential data. The results show that the real one, an incompletely connected market, was found to be more resilient to contagion than the complete one obtained by the maximum entropy.

In fact, the Mistrulli (2010) test only compares one actual matrix with the theoretical one, but the possible incomplete matrices can be really different among them, and the actual resilience to contagion also depends on the shock dimension.

The theoretical models by Allen and Gale (2000) and Freixas et al. (2000) are just a first approximation but of great importance with reference to possible regulation of the interbank lending. It was also an important reference on what was expected to happen when, as in some moments of the last financial crisis, the low confidence in other banks' stability has prompted the use of central banks as money centers (Table 2.19).

However, these network analyses on simple and clear structures of the interbank market (and matrix) are not actually useful for providing insights into the behavior of more complex systems, where different patterns are mixed up.

Memmel et al. (2012) analyzed the linkages between 14 large and internationally active German banks and the savings and cooperative sector, and obtained a 16×16 matrix of interbank exposures. They found for this sample an almost complete network, where only two off-diagonal elements were 0, that is, only two of the 240 possible interbank relations were not active.

Table 2.19 Money center structure.

	Bank 1	Bank 2	Bank 3	Bank 4	Bank 5		IB*d*
Bank 1	0	7	8	4	7		26
Bank 2	4	0	0	0	0		4
Bank 3	7	0	0	0	0		7
Bank 4	7	0	0	0	0		7
Bank 5	4	0	0	0	0		4
							48
IB*c*	22	7	8	4	7	48	

Craig and von Peter (2014) analyzed the whole German interbank network and found that out of 2182 banks, 1802 are active: 1671 as intermediaries, 67 as lenders only, and 64 as borrowers only.

These two analyses are not in contrast as it seems. In fact, the most widespread structure of the interbank market is of a multiple money center, also called a tiered structure.

This structure is characterized by a relatively small group of core banks, interconnected and each one also acting as a reference for some peripheral banks (see Figure 2.15). This allows the core banks to act in a well-diversified market with all the other core banks, and the peripheral banks to have one single (or a few) counterparts for their interbank lending and borrowing as in Table 2.20. If we limit the analysis to the core banks, the interbank market is (almost) complete, whereas if we consider all the banks, the matrix is typically incomplete but connected.

In fact, the actual patterns are not only more complex but also almost always not known, as banks in their balance sheets only declare total interbank credits and debts, and no information is provided as to what the counterparts are. So, the main problem is how to overcome the lack of data.

Another problem that has to be considered is the high volatility (see Figure 2.16) of the interbank exposures (see,

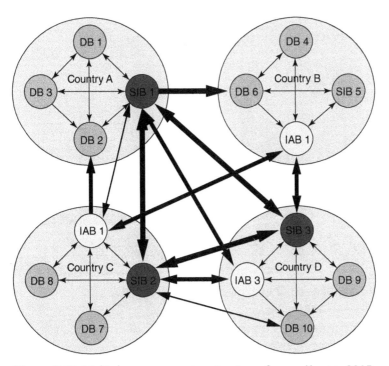

Figure 2.15 Multiple money center structure. *Source:* Kanno, 2015. Reproduced with permission of Elsevier.

for example, Garratt *et al.* (2011) and Gabrieli (2011)) that are mainly used for liquidity management, while the actual values of exposures are read from balance sheet data, annually or, in the best cases, quarterly.

Table 2.20 Multiple money center—tiered structure.

	Bank 1	Bank 2	Bank 3	Bank 4	Bank 5		IB*d*
Bank 1	0	6	3	0	2		11
Bank 2	5	0	0	3	0		8
Bank 3	0	3	0	0	0		3
Bank 4	0	0	0	0	0		0
Bank 5	4	0	0	0	0		4
							26
IB*c*	9	9	3	3	2	26	

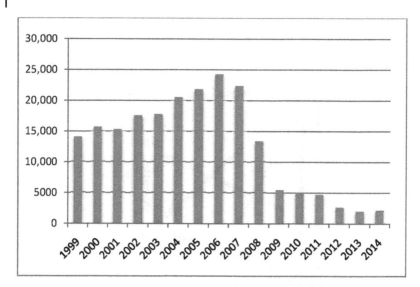

Figure 2.16 Interbank lending in Italy. *Source:* Banca d'Italia.

The two examples in Figure 2.17 and Table 2.20 give an insight into the actual use of the interbank exposures by the maturity distribution. In fact, about two thirds of the interbank exposures are on overnight contracts, which means that the actual

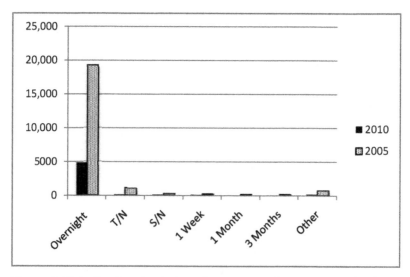

Figure 2.17 Italian interbank loans by maturity. *Source:* Banca d'Italia.

composition and volume of the exposures change for about the same share every day.

Some papers, such as Allen *et al.* (2012), analyzed the linkages between maturity and systemic risk, suggesting that the debt maturity has an important role and can affect contagion differently, depending on the interbank completeness.

In any case, the maturity distribution of interbank loans reveals a number of important issues.

One aspect is related to the estimation of important exposures for contagion. As the actual default of a bank is not so immediate, the counterpart banks have the possibility to not renew the overnight loans, so the only risky interbank loans are the ones with longer terms. The problem is in actually finding data on the specific maturity of single banks' exposures, since if the only relevant information available is on the average incidence for the whole system, this only results in reducing the LGD.

Another issue refers to the importance of the interbank balances for the dynamic liquidity management of banks, witnessed by the incidence of short-term exposures, and to the lowering of confidence after the 2008 crisis that led to a reduction in interbank exposures, which in Italy (Figure 2.17) dropped in 2010 to a value near a quarter of the value registered in 2005.

Something similar is in the Fedwire data (Table 2.21), which dropped from 491.4 during the early crisis to 368.9 when the crisis was easing, mainly due to a reduction in the domestic lending (from 389 to 271), while no great impact is recorded for foreign exposures, typically linked to risk diversification.

It is evident that when dealing with international samples, the maximum entropy hypothesis must be considered with greater attention. In fact, the tiered interbank structure allows this hypothesis to be used as an acceptable approximation with reference to the top-tier banks, those actually active in the international market and of larger dimensions (group 1 banks, EBA panel, etc.), while it is not acceptable with reference to small banks. Using the maximum entropy hypothesis on an international sample including banks of both categories will

Table 2.21 Maturity composition of Fedwire-settled interbank loans.

	Time period				
	Full sample	Precrisis	Early crisis	Crisis peak	Crisis easing
Dates	1/1/2007–3/31/2009	1/1/2007–8/9/2007	8/9/2007–9/12/2008	9/15/2008–11/11/2008	11/12/2008–3/31/2009
Overnight	300	279	330	302	248
2–7 days	17	16	17	15	20
8–14 days	13	11	13	14	14
>14 days but <1 month One month	18	14	19	20	24
1 month	21	20	26	9	14
2 months	8	9	9	4	6
3 months	24	31	28	8	11
4–5 months	12	15	13	11	6
6 months	12	12	14	9	7
7–11 months	16	17	16	12	15
1 year	6	8	6	3	5
Total	448	431	491	406	369

Source: Kuo, 2013. Reproduced with permission of Federal Reserve Bank of New York Staff Reports.

have the effect of simulating contagion as if the interbank exposures were spread as in Table 2.15, knowing that the actual interbank exposure structure is that shown in Table 2.19.

This would result in biased estimations. It is quite evident that if the larger banks, actually connected internationally, have a capitalization level higher than the smaller ones, a domestic crisis can have small effects on the other countries if the interbank exposures are tiered. If, instead, the model simulates the international system behavior as if all banks were connected to all other banks (maximum diversification and no tiering), the domestic crisis will spread its effects also in the other countries considered.

So, running simulations based on the maximum entropy hypothesis when the actual system interbank market is incomplete or tiered would be somewhat similar to setting a higher correlation than it actually is: Each bank default will affect all banks in the system (common shock), instead of only hitting the exposed ones, as is the case for idiosyncratic shocks.

Instead, it is possible to integrate other information sources in order to obtain a better approximation of the actual interbank matrix, such as the EBA disclosures on the geographical breakdown of individual bank activities for the eurozone and the BIS data on the international interbank claims (the values for a sample of countries are reported in Table 2.22).

Table 2.22 shows that similar to what happens for the interbank market for single banks, the international exposures between banking systems are also important and are of different types. While in some cases it seems to be due to rebalancing funding needs and availability (e.g., France versus Italy, the exposures being clearly unbalanced), in other cases it is possibly due to risk diversification (e.g., France versus Canada, where the

Table 2.22 International interbank claims on ultimate risk basis—Q4 2015, values in millions of US dollars.

	Canada	France	Germany	Italy	Japan	Switzerland	United Kingdom	United States
Canada		8,248	2,692	31	3,628	1,663	14,076	30,658
France	7,205		37,543	40,547	31,239	8,296	101,534	69,214
Germany	13,037	66,655		11,517	1,464	16,905	102,758	60,090
Italy	855	16,272	29,308		667	1,185	15,175	6,486
Japan	17,352	32,718	21,991	2,058		8,606	34,615	99,924
Switzerland	8,070	20,810	32,046	3,993	. . .		47,401	70,444
United Kingdom	14,884	44,164	38,839	3,531	34,887	6,966		72,336
United States	23,639	43,203	25,105	12,748	94,249	8,901	45,879	

Source: Reproduced with permission of BIS data wharehouse.

exposures are of similar value), similar to what happens for single banks.

Even if the matrix is almost complete, the same maximum entropy hypothesis is not actually respected (e.g., the exposure of Italy to Germany is of about 40% of the country total, 29 out of 72 billion dollars, while the system is exposed to Germany for only 12%), so including this information in the matrix estimation adds important information, thus improving the estimation reliability.

2.9.1 Interbank Matrix Estimation

In mathematical terms, when dealing with a system of n banks, we have to define a matrix of $n \times n$ elements

$$
IB = \begin{pmatrix}
IB_{1,1} & IB_{1,2} & \cdots & IB_{1,n-1} & IB_{1,n} \\
IB_{2,1} & IB_{2,2} & \cdots & IB_{2,n-1} & IB_{2,n} \\
\vdots & \vdots & \ddots & \vdots & \vdots \\
IB_{n-1,1} & IB_{n-1,2} & \cdots & IB_{n-1,n-1} & IB_{n-1,n} \\
IB_{n,1} & IB_{n,2} & \cdots & IB_{n,n-1} & IB_{n,n}
\end{pmatrix}
$$

$$(2.32)$$

only knowing these values:

$\sum_j IB_{ji}$ total of interbank credits of bank i, also written as IBc_i, and this for all banks $i = 1, \ldots, n$,

$\sum_j IB_{ij}$ total of interbank debts of bank i, also written as IBd_i, for all banks $i = 1, \ldots, n$,

i.e. only having the row totals and column totals of the IB matrix.

In this simple framework, each element can be estimated as

$$
IB_{i,j} = \frac{IBd_j}{\sum_k IBd_k} IBc_i \tag{2.33}
$$

i.e. the interbank debt of bank i versus bank j is given by the share of the system total debts of bank j multiplied by the value of interbank credits of the system held by bank i.

In fact, one more implicit constraint is to be considered, as banks do not lend to themselves, so

$$IB_{ji} = 0 \forall i = j$$

or all the diagonal elements of the matrix are zero:

$$IB = \begin{pmatrix} 0 & IB_{1,2} & \cdots & IB_{1,n-1} & IB_{1,n} \\ IB_{2,1} & 0 & \cdots & IB_{2,n-1} & IB_{2,n} \\ \vdots & \vdots & \ddots & \vdots & \vdots \\ IB_{n-1,1} & IB_{n-1,2} & \cdots & 0 & IB_{n-1,n} \\ IB_{n,1} & IB_{n,2} & \cdots & IB_{n,n-1} & 0 \end{pmatrix}$$

$$(2.34)$$

If no more information is available on the interbank exposure, we now have to find some rule for estimating the exposure of each bank to each of the others, and fill the IB matrix with it.

The most common hypothesis is that banks tend to diversify risk sources, so to spread exposures on all possible counterparts, also known as the maximum entropy hypothesis.

Some studies analyzed the actuality and the impact of this hypothesis on contagion estimation, such as Mistrulli (2010) and Zedda *et al.* (2012a). Another interesting robustness test is developed in Hałaj and Kok (2013), which, in considering the high volatility of the interbank exposures, verified whether variations in the interbank exposures induce important variations in results or whether the estimations are substantially stable.

Their approach and results are presented in the next section.

In this framework, it must be that summing up row totals (interbank credits) or summing up column totals (interbank debts) yields the same result:

$$\sum_j IBc_j = \sum_j IBd_j$$

When considering actual data, we normally deal with a sample of banks, so that the interbank exposures are not balanced. In fact, even when dealing with a whole national banking system, the international interbank lending is not generally balanced.

To overcome this problem, we have to add a row and a column to the matrix, representing the net exposures to the "rest of the world":

$$IB = \begin{pmatrix} 0 & IB_{1,2} & \cdots & IB_{1,n} & IB_{1,n+1} \\ IB_{2,1} & 0 & \cdots & IB_{2,n} & IB_{2,n+1} \\ \vdots & \vdots & \ddots & \vdots & \vdots \\ IB_{n,1} & IB_{n,2} & \cdots & 0 & IB_{n,n+1} \\ IB_{n+1,1} & IB_{n+1,2} & \cdots & IB_{n+1,n} & 0 \end{pmatrix}$$

(2.35)

This difference can be attributed to banks in proportion to their total exposure as follows:

$$\sum_i \sum_j IB_{ji} > \sum_j \sum_i IB_{ij}$$

The last row will be of zeros, and the last column will contain the values balancing the higher debt, so that the interbank matrix becomes as follows:

$$IB = \begin{pmatrix} 0 & IB_{1,2} & \cdots & IB_{1,n} & IB_{1,n+1} \\ IB_{2,1} & 0 & \cdots & IB_{2,n} & IB_{2,n+1} \\ \vdots & \vdots & \ddots & \vdots & \vdots \\ IB_{n,1} & IB_{n,2} & \cdots & 0 & IB_{n,n+1} \\ 0 & 0 & \cdots & 0 & 0 \end{pmatrix}$$

(2.36)

while $\sum_i \sum_j IB_{ij} < \sum_j \sum_i IB_{ij}$

The last column will only contain zeros, and the last row will contain the values balancing the net credit of the sample. The matrix will be as follows:

$$IB = \begin{pmatrix} 0 & IB_{1,2} & \cdots & IB_{1,n} & 0 \\ IB_{2,1} & 0 & \cdots & IB_{2,n} & 0 \\ \vdots & \vdots & \ddots & \vdots & \vdots \\ IB_{n,1} & IB_{n,2} & \cdots & 0 & 0 \\ IB_{n+1,1} & IB_{n+1,2} & \cdots & IB_{n+1,n} & 0 \end{pmatrix}$$

(2.37)

Finally, we have to respect the constraints on row totals, column totals, zeros on diagonal, maximum diversification, and possible unbalanced values. Putting everything together, we have

$$
I\hat{B} = \begin{pmatrix}
0 & I\hat{B}_{1,2} & \cdots & I\hat{B}_{1,n} & I\hat{B}_{1,n+1} \\
I\hat{B}_{2,1} & 0 & \cdots & I\hat{B}_{2,n} & I\hat{B}_{2,n+1} \\
\vdots & \vdots & \ddots & \vdots & \vdots \\
I\hat{B}_{n,1} & I\hat{B}_{n,2} & \cdots & 0 & I\hat{B}_{n,n+1} \\
I\hat{B}_{n+1,1} & I\hat{B}_{n+1,2} & \cdots & I\hat{B}_{n+1,n} & 0
\end{pmatrix}
$$

(2.38)

where all $I\hat{B}_{ij}$ elements are to be computed by means of a constrained entropy maximization algorithm, as for the RAS by Blien and Graef.

Example: A five-bank system

One example can provide a better explanation for how the process is developed.

In balance sheets, we only have the total interbank exposure values as in Table 2.23.

Directly computing the interbank matrix by means of the maximum entropy hypothesis will give the values in Table 2.24.

Here the $IB_{1,2}$ element is given by $(13/36) \times 8$, $13/36$ being the share of the system total debts held by 2, and 8 the interbank debt of bank 1 to be shared.

But if we consider that banks do not lend money to themselves, we have to input zeros on diagonal elements. However,

Table 2.23 Total interbank exposures.

	IB*d*	IB*c*
Bank 1	8	1
Bank 2	6	13
Bank 3	12	7
Bank 4	7	3
Bank 5	3	12
Total	36	36

Table 2.24 Interbank matrix maximum diversification.

	Bank 1	Bank 2	Bank 3	Bank 4	Bank 5		Row sum	IBd
Bank 1	0.22	2.89	1.56	0.67	2.67		8.00	8
Bank 2	0.17	2.17	1.17	0.50	2.00		6.00	6
Bank 3	0.33	4.33	2.33	1.00	4.00		12.00	12
Bank 4	0.19	2.53	1.36	0.58	2.33		7.00	7
Bank 5	0.08	1.08	0.58	0.25	1.00		3.00	3
							36.00	36
Column sum	1.00	13.00	7.00	3.00	12.00	36.00		
IBc	1	13	7	3	12	36		

just substituting zeros on the diagonal elements of the previous matrix (as in Table 2.25) is not sufficient, as the total no longer holds.

Thus, we have to distribute the values formerly attributed to the diagonal to the other banks. If we share this additional value among the other banks in the same row (Table 2.26), the row totals will now hold:

Table 2.25 Interbank matrix estimation: inserting zero on diagonal.

	Bank 1	Bank 2	Bank 3	Bank 4	Bank 5		Row sum	IBd
Bank 1	0	2.89	1.56	0.67	2.67		7.78	8
Bank 2	0.17	0	1.17	0.50	2.00		3.83	6
Bank 3	0.33	4.33	0	1.00	4.00		9.67	12
Bank 4	0.19	2.53	1.36	0	2.33		6.42	7
Bank 5	0.08	1.08	0.58	0.25	0		2.00	3
							29.69	36
Column sum	0.78	10.83	4.67	2.42	11.00	29.69		
IBc	1	13	7	3	12	36		

Table 2.26 Interbank matrix estimation, rebalancing on rows.

	Bank 1	Bank 2	Bank 3	Bank 4	Bank 5		Row sum	IBd
Bank 1	0.00	2.97	1.60	0.69	2.74		8.00	8
Bank 2	0.26	0.00	1.83	0.78	3.13		6.00	6
Bank 3	0.41	5.38	0.00	1.24	4.97		12.00	12
Bank 4	0.21	2.76	1.48	0.00	2.55		7.00	7
Bank 5	0.13	1.63	0.88	0.38	0.00		3.00	3
							36.00	36
Column sum	1.01	12.73	5.79	3.08	13.38	36.00		
IBc	1	13	7	3	12	36		

Here the $IB_{1,2}$ element is given by $(13/(36-1)) \times 8$, where the interbank debt of bank 1, with a value of 8, is shared by all banks but bank 1, and the share of the system total debts held by bank 2 is $13/(36-1)$.

Some papers in fact use this way for sharing the losses to the counterpart banks in proportion to their exposure to the defaulted bank.

From the table, we can see that row totals hold, and the diagonal zeros constraint is maintained. But column totals are not matching IBc values!

This explains why we have to use a specific algorithm to correctly compute the values of each exposure.

The most cited and used in this framework is the RAS algorithm developed by Blien and Graef (1997).

The method consists in repeating the same approach used above for distributing the difference between the row total and the actual interbank debt, alternating one step working on rows and one step working on columns, and so on up to convergence.

Continuing from the previous example, the next step is in computing the columns' differences and sharing them among the banks in the same column as in Table 2.27.

Table 2.27 Interbank matrix estimation, rebalancing on columns.

	Bank 1	Bank 2	Bank 3	Bank 4	Bank 5		Row sum	IB*d*
Bank 1	0.00	3.04	2.00	0.66	2.41		8.12	8
Bank 2	0.26	0.00	2.13	0.77	2.88		6.03	6
Bank 3	0.41	5.49	0.00	1.21	4.46		11.56	12
Bank 4	0.21	2.82	1.84	0.00	2.25		7.12	7
Bank 5	0.12	1.65	1.03	0.37	0.00		3.17	3
							36.00	36
Column sum	1.00	13.00	7.00	3.00	12.00	36.00		
IB*c*	1	13	7	3	12	36		

This way the column totals match, but the row totals no longer hold. However, the approximation in row totals has improved from the previous step, so now the differences between the actual row total and the target one are lower than 0.1 for all rows. Continuing further, the procedure converges to a table where both row totals and column totals eventually match the actual value of interbank debts and interbank credits (Table 2.28).

Table 2.28 Interbank matrix estimation, final values.

	Bank 1	Bank 2	Bank 3	Bank 4	Bank 5		Row sum	IB*d*
Bank 1	0.00	2.99	2.02	0.65	2.34		8.00	8
Bank 2	0.26	0.00	2.15	0.76	2.84		6.00	6
Bank 3	0.42	5.69	0.00	1.25	4.63		12.00	12
Bank 4	0.20	2.76	1.85	0.00	2.19		7.00	7
Bank 5	0.11	1.55	0.99	0.34	0.00		3.00	3
							36.00	36
Column sum	1.00	13.00	7.00	3.00	12.00	36.00		
IB*c*	1	13	7	3	12	36		

Table 2.29 Unbalanced interbank exposures.

	IB*d*	IB*c*
Bank 1	10	2
Bank 2	6	12
Bank 3	8	5
Bank 4	20	4
Bank 5	5	17
Total	49	40

When dealing with actual samples or systems, we often find that total interbank credits and total interbank debts do not match, as in the example in Table 2.29.

Here the total interbank credits are 40, while the total interbank debts are 49.

The solution is really simple and involves adding a line to represent the rest of the world (RoW) and using this line to balance the values, as shown in Table 2.30.

We can then compute the matrix as before, for a system with six banks, as in Table 2.31.

In many of the studies cited above, other sources of information were considered. The sources of this integrative information were of different kinds.

Table 2.30 Interbank exposure integration.

	IB*d*	IB*c*
Bank 1	10	2
Bank 2	6	12
Bank 3	8	5
Bank 4	20	4
Bank 5	5	17
RoW	0	9
Total	49	49

Table 2.31 Interbank matrix with rest-of-the-world (RoW) maximum entropy.

	Bank 1	Bank 2	Bank 3	Bank 4	Bank 5	RoW	Row sum	IBd
Bank 1	0.00	0.34	0.38	0.96	0.32	0	2.00	2
Bank 2	2.52	0.00	2.16	5.48	1.84	0	12.00	12
Bank 3	1.08	0.81	0.00	2.33	0.78	0	5.00	5
Bank 4	1.25	0.88	1.05	0.00	0.82	0	4.00	4
Bank 5	3.55	2.70	3.04	7.71	0.00	0	17.00	17
RoW	1.60	1.26	1.38	3.52	1.24	0	9.00	9
Column sum	10.00	6.00	8.00	20.00	5.00	0.00	49.00	49
IBc	10	6	8	20	5	0	49	

An interesting example is that of Elsinger *et al.* (2006a), where the particular structure of the Austrian banking system and reporting standards helped with the estimation of the matrix, allowing 72% of the matrix entries. In some other cases, data are available on interbank large exposures, as in Degryse and Nguyen (2007), who had access to bank data reports including interbank exposures exceeding 10% of their own funds. Another source of information is the international interbank claims published by the BIS for a sample of countries.

Generally speaking, we have to deal with two possible kinds of additional information, either exact values for some element of the matrix or values for a subgroup of counterpart banks.

While the first kind of information sounds simpler to be included, just filling in the value in the matrix, in fact it results in changing the total to be distributed on the remaining banks, thus a constraint on the total for a subgroup of counterpart banks.

This inclusion can be managed in the same way as above after considering that no bank lends to itself, and filling in the diagonal elements to zero. In fact, the only difference is in the value, zero in the previous case and a positive amount in this case.

Table 2.32 Netted Interbank matrix.

	Bank 1	Bank 2	Bank 3	Bank 4	Bank 5	RoW	Row sum	IBd
Bank 1	0	0	0	0	0	0	0.00	2
Bank 2	2.19	0	1.34	4.59	0	0	8.12	12
Bank 3	0.70	0	0	1.28	0	0	1.98	5
Bank 4	0.29	0	0	0	0	0	0.29	4
Bank 5	3.23	0.86	2.26	6.89	0	0	13.24	17
RoW	1.60	1.26	1.38	3.52	1.24	0	9.00	9
Column sum	8.01	2.12	4.98	16.28	1.24	0	32.63	49
IBc	10	6	8	20	5	0	49	

This integration of different sources, i.e. a constrained maximization of the matrix entropy, can be solved using the same computational tools cited above.

Degryse and Nguyen (2007) also performed contagion simulations based on "netted" matrices of domestic bilateral exposures (Table 2.32).

These simulations assume that all the interbank claims are covered by bilateral netting agreements. They found that netting substantially reduces contagion toward very low levels. This result, however, can be affected by the baseline scenario, a complete interbank matrix, while possibly the real cases where a bank is at the same time creditor and debtor of the same bank may in practice occur in a limited number of cases.

2.9.2 Robustness Checks on the Maximum Entropy Hypothesis

The maximum entropy hypothesis is an important tool as it allows the interbank matrix to be completed, even in presence of limited information. However, it is important to determine what the impact of this hypothesis is on estimations.

In the literature, different types of testing were developed in order to have some more references on the actual concentration of interbank exposures (see Mistrulli, 2010) and verify how different concentration levels affect the estimations.

The work developed by Mistrulli (2010), based on confidential information on the actual interbank exposures from the central bank, evaluated the difference between contagion simulations results based on the maximum entropy hypothesis and on the actual interbank matrix.

Results showed that in the case at hand, with an incomplete interbank market, the maximum entropy method leads to lower values for the financial contagion risks than that which resulted when considering the actual interbank exposures.

On the other hand, a theoretical test on the effects of concentration was developed in Zedda et al. (2012a).

In this paper, variations in the interbank matrix were induced to test the robustness of the maximum entropy assumption by evaluating if these changes produce a significant variation in results.

Starting on the matrix of bilateral interbank exposures obtained via the ME hypothesis, the variations are introduced with a procedure that preserves totals but introduces one zero more at each step. In this way an incomplete matrix is obtained that concentrates interbank exposures to a limited preset number of nonzero values.

The procedure develops as follows: Considering, for example, a 5×5 IB matrix,

$$
IB_{5 \times 5} = \begin{pmatrix}
0 & x_{12} & x_{13} & x_{14} & x_{15} \\
x_{21} & 0 & x_{23} & x_{24} & x_{25} \\
x_{31} & x_{32} & 0 & x_{34} & x_{35} \\
x_{41} & x_{42} & x_{43} & 0 & x_{45} \\
x_{51} & x_{52} & x_{53} & x_{54} & 0
\end{pmatrix}
$$

1) Select, randomly, two different rows and two different columns, which identify four different elements of the matrix (e.g., lines 1 and 2 and columns 3 and 4 identifies $x_{1,3}$, $x_{1,4}$, $x_{2,3}$, $x_{2,4}$). Provided all four values are different from zero,

these elements are going to be changed in order to obtain a new matrix with one additional zero:

$$IB_{5\times5} = \begin{pmatrix} 0 & x_{12} & 5 & 8 & x_{15} \\ x_{21} & 0 & 6 & 12 & x_{25} \\ x_{31} & x_{32} & 0 & x_{34} & x_{35} \\ x_{41} & x_{42} & x_{43} & 0 & x_{45} \\ x_{51} & x_{52} & x_{53} & x_{54} & 0 \end{pmatrix} \qquad (2.39)$$

2) Evaluate which one of the four elements has the lowest value (in the example the lowest value is the element $x_{1,3} = €5$ trillion).
3) The lowest value is subtracted to itself and to the element in the other row and other column $x'_{13} = x_{13} - 5 = 0; x'_{24} = x_{24} - 5$ and added to the element with the same row but different column and the same column but different row $x'_{23} = x_{23} + 5; x'_{14} = x_{14} + 5$.

The new matrix IB′ will be

$$IB'_{5\times5} = \begin{pmatrix} 0 & x_{12} & 5-5 & 8+5 & x_{15} \\ x_{21} & 0 & 6+5 & 12-5 & x_{25} \\ x_{31} & x_{32} & 0 & x_{34} & x_{35} \\ x_{41} & x_{42} & x_{43} & 0 & x_{45} \\ x_{51} & x_{52} & x_{53} & x_{54} & 0 \end{pmatrix} \qquad (2.40)$$

$$\Rightarrow IB'_{5\times5} = \begin{pmatrix} 0 & x_{12} & 0 & 13 & x_{15} \\ x_{21} & 0 & 11 & 7 & x_{25} \\ x_{31} & x_{32} & 0 & x_{34} & x_{35} \\ x_{41} & x_{42} & x_3 & 0 & x_{45} \\ x_{51} & x_{52} & x_{53} & x_{54} & 0 \end{pmatrix} \qquad (2.41)$$

In this way row and column totals are respected, but a zero is introduced where the lowest value was originally placed. This procedure is then iterated, up to the preset number of zeros.

For each country, starting from the maximum entropy, the matrix was modified to obtain 20 series of interbank matrices

Table 2.33 Variability—average value in standard error of contagion simulations results.

	+20% zeros	+35% zeros	+50% zeros	+65% zeros	+80% zeros
BE	0.6%	0.7%	2.1%	3.3%	5.7%
IE	11%	26%	34%	56%	78%
IT	0.004%	0.008%	0.013%	0.023%	0.045%
PT	2%	4%	4%	6%	7%

Source: Zedda, 2012. Reproduced with permission of European Union.

with 20, 35, 50, 65, and 80% more elements set to zero (other than diagonal elements or elements already set at zero).

To perform a "ceteris paribus" analysis, for each simulation the variations in interbank matrix are set randomly, while the internal losses suffered by each bank (see next section) are always the same. In this way, different results for the same country can only be due to variations in the interbank matrix.

Table 2.33 reports the first expected effect of the higher concentration of interbank exposures: the higher variability. In fact, while the maximum entropy can only have one possible solution, so its values are determined by the totals and zero diagonal constraint; higher concentration can be the result of different possible realizations of the same totals, and the higher the concentration, the higher the possibility of different solutions (degrees of freedom). In this exercise, the process is even more evident, as the higher concentration is obtained iterating the concentration loop up to the target concentration, and so a higher concentration results from more iterations of the same process. As each iteration is set randomly, each one has a different concentration, so there are more different possibilities to exploit.

From this point of view, the Mistrulli (2010) test can be considered as being based on one possible value of the interbank matrix with a higher concentration, reporting the effects of those particular values. Instead, the actual assessment of the role of concentration on contagion needs a broader investigation on different samples and settings.

Table 2.34 Average losses distribution by scenario—Ireland.

	Mean	1st–3rd quartile range	10th–90th quantile range
No contagion	1,707,946		
Base	16,998,231		
+20% zeros	17,049,103	2.6%	11.7%
+35% zeros	16,968,441	10.1%	32.9%
+50% zeros	17,206,620	26.2%	53.1%
+65% zeros	17,321,954	41.6%	78.3%
+80% zeros	19,941,129	71.2%	116.5%

Source: Zedda, 2014. Reproduced with permission of Springer.

The effect is more visible considering only one country, as in Table 2.34, which reports the interquartile range obtained for Ireland, as shown graphically in Figure 2.18.

The values in Table 2.34 show that the mean value is only slightly affected by the different levels of interbank exposures concentration, with some evident effect on the mean value only for the maximum value of concentration.

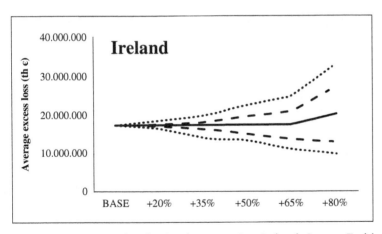

Figure 2.18 Losses distribution by scenario—Ireland. *Source:* Zedda, 2014. Reproduced with permission of Springer.

Table 2.35 Number of defaults by contagion magnitude—Ireland.

	Overall	Contagion		
	(10,000 cases)	No (6,174 cases)	Small (937 cases)	Large (2,889 cases)
Base	4.41	1.00	2.21	12.40
+20%	4.46	1.03	2.41	12.45
+35%	4.51	1.11	2.81	12.35
+50%	4.65	1.19	3.73	12.34
+65%	4.82	1.49	4.30	12.10
+80%	5.51	2.34	6.58	11.94

Source: Zedda, 2012. Reproduced with permission of European Union.

The analysis of the same results on the basis of contagion magnitude reveals important details on the effect of concentration.

Selecting the simulations where, in the maximum entropy (base) case, contagion was not present, limited, or important, based on a comparison of the outcome of simulations with contagion and without contagion, we have different results for each category (Tables 2.35 and 2.36, and Figure 2.19).

Table 2.36 Average value of losses by contagion magnitude—Ireland.

	Overall	Contagion		
	(10,000 cases)	No (6,174 cases)	Small (937 cases)	Large (2,889 cases)
Base	16,998,231	989,367	2,394,299	55,946,867
+20%	17,049,103	1,074,374	3,407,988	55,612,516
+35%	16,968,441	1,291,853	5,529,794	54,180,373
+50%	17,206,620	1,565,217	9,522,706	53,125,572
+65%	17,321,954	2,517,032	12,076,994	50,662,249
+80%	19,941,129	6,014,341	22,203,313	48,969,972

Source: Zedda, 2012. Reproduced with permission of European Union.

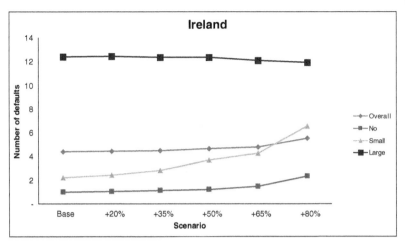

Figure 2.19 Number of defaults by contagion magnitude—Ireland.
Source: Zedda, 2012. Reproduced with permission of European Union.

The no contagion and low contagion cases suggest an under-estimation of contagion, in terms of both the number of defaults and the value, for the estimations based on the maximum entropy value for the interbank matrix. Instead, for large crises, maximum entropy seems to overestimate contagion effects, as found by Mistrulli (2010).

Summing up, the results reported above, based on a sample of 20 interbank matrices for each level of concentration, already suggest that different values for capitalization, interbank exposures incidence and assets riskiness, different crisis magnitude, and possibly other significant values such as correlation affect results in such a complex way that it is not evident whether the concentration has a linear impact on the estimates. More detailed analyses and tests will possibly help.

Another important contribution is that of Hałaj and Kok (2013). Their approach differs from networks based on entropy methods and real-time data, which typically capture one particular snapshot of the network structure (Figure 2.20).

The method is based on the simulation of a large number of possible networks, coherently with the underlying exposure data and imposed behavioral characteristics. This produces a very dynamic pattern of interbank networks, which reflects the

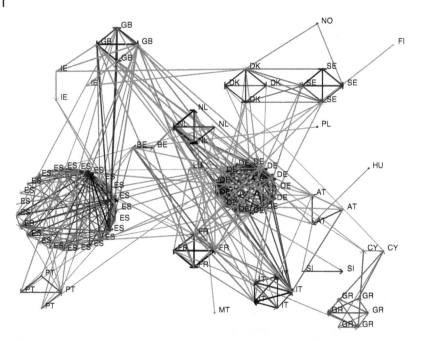

Figure 2.20 European banking network. *Source:* Halai, 2013.
Reproduced with permission of European Central Bank.

volatile nature of financial network structures as confirmed by
many studies (see, for example, Garratt *et al.* (2011); Gabrieli
(2011)).

Starting from the EBA disclosures on the geographical breakdown of individual bank activities, aggregated by country, the
fraction of the exposures toward banks in each country was
calculated.

These fractions were assumed to be probabilities that a bank in
a given country makes an interbank placement to a bank in
another (or the same) country. Banks were grouped into two
subcategories within countries: with domestic scope of activities
and with international activity, respectively.

Instead of starting from the maximum entropy (as in Zedda
et al., 2012a), the network is generated randomly, based on the
probability map, starting from the empty matrix. At each step a
pair of banks is randomly drawn, and a random fraction of the
reported interbank liabilities is assigned to the linkage. Then,

the stock of interbank liabilities and assets is reduced by the volume of the assigned placement, and the procedure is repeated until no more interbank liabilities are left.

By this procedure, they construct a sample of 20,000 simulated interbank networks.

Here also, comparing the distribution of the obtained results, and in particular the maximum entropy network and the average of results obtained by the simulated networks, it is suggested that no major differences appear in the point estimation of expected results.

Instead, from the richness of the sample considered, one can verify that contagion is heterogeneous across the banking system and strongly nonlinear, so a small fraction of possible network structures may spread important contagion losses across the system.

2.10 Loss Given Default

The estimation of the LGD banks is difficult, since, fortunately, bank failures rarely happen. Moreover, actual losses on a defaulting bank can prove very complicated to calculate, since they depend on the time horizon chosen.

In fact, the first effect of counterpart default is in the immediate imbalance in financial flows that induce the need for rebalancing by reducing the new loans issue or selling some asset. In a longer horizon, the default will be almost partially recovered by the fraction of residual assets of the defaulted firm, and some value re-entering the bank balance sheet. As this process takes time, it is not simple to measure the actual effect of defaults, as the recovery is typically acknowledged several years later from the default, and so the effect cannot be directly related to the cause.

Instead, the liquidity shock can be considered instantaneous, and the effect can be proxied by the 100% LGD when simulating liquidity shocks.

The problem is similar when dealing with interbank defaults, as having a bank as counterpart in any case has the same effect in destabilizing the financial planning.

Apart from the difficulties in estimating the LGD, more complexity is added by its variability over time. The evaluations in the studies of Bruche and González-Aguado (2010) and Altman *et al.* (2005) suggest that default rates and average recovery rates are negatively correlated, both being driven by the same common factor clearly related to the business cycle; so in recessions or industry downturns, default rates are high, and recovery rates are low.

The fact that recovery rates are low when defaults are frequent is important, because this amplifies systematic risk. It may thus be advisable to model the LGD as a function of the level of distress in the financial sector.

2.10.1 Constant LGD

Altman and Vellore (1996) estimated average recovery rates on defaulting bonds of financial institutions (for the period 1978–1995) ranging, by type of institution, from 68% for mortgage banks to 9% for savings institutions and being about 36% on average. But the LGD for bonds is probably very different from the LGD for comparable loans (which include secured and unsecured assets).

James (1991) estimated the impact of interbank failures in terms of loss given default distinguishing between direct and indirect costs and found that the direct losses average 30% of the exposure. At default, the expenses associated with bank closures average 10% of the EAD, summing up to a total of about 40% of the EAD.

Degryse and Nguyen (2007), on the basis of the actual kind of interbank exposures and knowing that more than 50% of interbank loans granted by Belgian banks are secured, estimated that to be realistic, for the specific case, one should assume a recovery between 60 and 80%, thus an LGD between 40 and 20%.

Memmel *et al.* (2012) analyzed the German market by means of a simulation model and estimated the LGD of interbank exposures between 1990 and 2008 (see Table 2.37).

The effect of LGD values is nevertheless important and interacts with the other variables and hypotheses.

Table 2.37 Means and standard deviations of the empirical frequency distribution of the LGD on different samples of lender banks.

Sample	N	Mean	Standard deviation
All banks	667	0.38	0.39
Large and internationally active banks	101	0.45	0.32
Savings banks	50	0.58	0.38
Cooperative banks	222	0.24	0.37

Source: Memmel, 2012. http://www.ijcb.org/copyright.htm

As an example, Mistrulli (2010) found that the observed bilateral exposures generate a higher degree of contagion (as a share of total assets) for low LGDs, whereas the opposite holds for large LGDs.

With reference to liquidity, as domino effects may be considered instantaneous, one could also argue that the time pattern of recovery does not matter and that an LGD of 100% should be used to simulate liquidity shocks. Yet, the time pattern of recovery may matter, depending on the maturity of the liabilities.

2.10.2 Stochastic LGD

A second possibility was explored by Memmel *et al.* (2012). In their analysis, based on an exogenous single failure model, they verified the effects on contagion when using a stochastic value for the LGD instead of a fixed value.

They analyzed the LGD values distribution for each banking category (see Figure 2.21) and generated LGD values according to a beta distribution with the same parameters estimated for the actual distribution of the considered category. Relative frequency was derived from data on German private commercial banks and the central institutions of the savings and cooperative banks. There were 344 observations for the period 1990–2008.

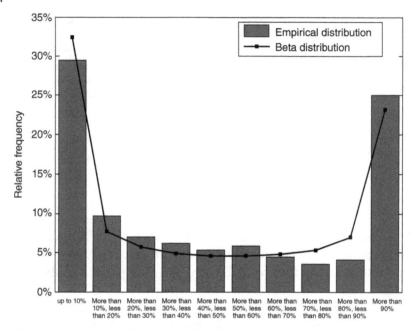

Figure 2.21 Relative frequency of the loss given default for interbank loans. *Source:* Memmel, 2012. http://www.ijcb.org/copyright.htm

Their results are from an average of 2.69 bank failures under the assumption of a constant LGD, while the average number of failures under the assumption of a stochastic LGD was 4.06. They concluded that there is a certain risk of underestimating the effects of a bank failure on financial stability if the distribution of the LGD is not considered.

2.10.3 Endogenous LGD

A third possibility is to model the actual excess loss for quantifying the final loss after a bank default episode. In this case, the model must include a more detailed representation of the contagion mechanism, including the actual initial losses but also the feedback losses induced to the initially failed bank by the other banks failed by contagion. This backward propagation effect induces more losses for the first step defaults, and so on in a loop, after which it is possible to estimate the actual final losses.

This modeling needs to include information on the actual debts of the distressed bank, the bankruptcy cost, and possibly the degree of collateralization.

Degryse and Nguyen (2007) developed this approach for endogenizing banks LGD. The first evidence is that it allows LGD to vary across banks.

In their model, to avoid feedback loop effects, the LGD is not considered for determining whether a bank defaults or not, but only to determine the losses in case of default.

The value of losses, thus of LGD, is determined on the basis of a system of equations, one for each bank, defined as follows:

$$\text{LGD}_i = \left[\frac{\sum_j \left(\text{LGD}_j x_{ij}\right) + \text{remaining assets} \times \text{LGD}_{ra}}{\text{Total assets} - \text{shareholders equity}} \right]$$

$$(2.42)$$

where the LGD on other assets is exogenous, but different for different assets classes, widely ranging from a 100% recovery rate ($\text{LGD} = 0$) for liquid assets to 100% of loss for intangible assets.

They found that the simulations reveal an evolution over time of contagion risk; however, at any given point in time, they no longer observe a strong correlation between the average implied LGD across banks and the level of contagion.

They argue this happens because the average LGD interacts with the other dimensions of the market structure, which remain determinant in the propagation of contagion. In addition, for a given average LGD across banks, contagion risk is higher when there is more cross-sectional variation in LGD.

Like Memmel *et al.* (2012), Degryse and Nguyen (2007) suggested that heterogeneity in LGD appears to exacerbate contagion risk.

Another possibility can be based on the Eisemberg and Noe algorithm (see Section 2.12), which is based on the computation of feedback effects at each step of contagion. The outcome of this algorithm, as the actual losses for each bank are computed also considering the effects of the induced defaults on the initially defaulted banks, tends to result in a higher LGD for the largest crises, and lower rates for smaller crises, in line with the findings of Bruche and González-Aguado (2010) and Altman *et al.* (2005), which suggest that default rates and average recovery rates are driven by the same common factor, so that during crises, default rates are high, and recovery rates are low.

This is also a way for implicitly modeling the LGD as a function of the level of distress in the financial sector.

2.11 Interbank Losses Attribution

After estimating the exposure at default, and the loss given default, the next step is to consider how to attribute the losses to the counterpart banks.

The simplest and more commonly used mechanism of losses attribution is simple proportional sharing by counterparts, on the basis of the EAD.

But from the theoretical point of view, if the losses are attributed to banks not in proportion to their exposures, but on the basis of their loss-absorbing capacity, the system can appear to be more stable, at least for limited crises.

The example in Table 2.38 can explain how it works.

Table 2.38 Interbank losses attribution on the basis of exposures share.

	Bank 1	Bank 2	Bank 3	Bank 4	Defaults	IB losses
IBc share	40%	20%	15%	25%		
Capital 1	10	−1	1	3	Bank 2	8
Losses 1	4		1.5	2.5		
Capital 2	6	−1	−0.5	0.5	Bank 3	5
Losses 2	2.4	1.2		1.5		
Capital 3	3.6	−2.2	−0.5	−1.0	Bank 4	3
Losses 3	1.6	0.8	0.6			
Capital 4	2.0	−3.0	−1.1	−1.0		

Table 2.39 Interbank losses attribution on the basis of loss-absorbing capacity.

	Bank 1	Bank 2	Bank 3	Bank 4	Defaults	IB losses
IBc share	40%	20%	15%	25%		
Capital 1	10	−1	1	3	Bank 2	8
Losses 1	5.71		0.57	1.71		
Capital 2	4.29	−1.00	0.43	1.29		

After the first shock, with bank 2 defaulting, losses are distributed on the basis of each bank exposures. As a result, bank 3 defaults in the second round, spreading more losses to the other banks, and making bank 4 default in the third round.

If, instead, we attribute losses based on the loss-absorbing capacity (as shown in Table 2.39), we find that the effect of the first default is more effective on banks with more residual capital so that no banks default in the second round.

This approach for interbank losses attribution has the effect of maximizing the loss-absorbing capacity of the system. From a different point of view, it sharpens the edge effect on contagion, since, when some spare loss-absorbing capacity is in the system, contagion is stopped, whereas if the losses exceed this threshold, the whole system will suddenly crash.

2.12 Contagion Simulation Methods

As soon as the estimation of the interbank exposures is solved, LGD is in some way quantified, losses are attributed to counterparts, and a contagion mechanism must be set or hypothesized for determining what happens after the first round of defaults.[1]

---

1 It is important to keep in mind that the iterations are merely a computational device; in principle, contagion is instantaneous.

Table 2.40 Interbank exposures at step 1.

	Bank 1	Bank 2	Bank 3	Total
Bank 1	0	0	4	4
Bank 2	6	0	2	8
Bank 3	6	2	0	8
Total	12	2	6	

In the literature, two different approaches were developed, the main distinction being in allowing or not allowing feedback effects on the already defaulted banks.

The first one, known as the Eisemberg and Noe fictitious default algorithm, is based on an iterative process where, after the first default (or defaults), if any other bank defaults by contagion, the losses on the interbank exposures are attributed even to the banks that failed in the previous step.

In Table 2.40, the starting values are those in Tables 2.41 and 2.42.

In this case, bank 2 has 8 of interbank debts, while only 2 of equity plus 2 of interbank credits. So bank 2 defaults and can pay only 4 out of 8 of its interbank debts. As bank 2 is exposed to bank 1 for 6 and to bank 3 for 2, their claims will be proportionally reduced to the available wealth, to 3 and 1.

So the new situation at step 2 will be as reported in Tables 2.42 and 2.43.

Table 2.41 Bank balances at step 1.

	Equity	IBd	IBc	Net
Bank 1	2	4	12	10
Bank 2	2	8	2	−4
Bank 3	2	8	6	0

Table 2.42 Interbank exposures at step 2.

	Bank 1	Bank 2	Bank 3	Total
Bank 1	0	0	4	4
Bank 2	3	0	1	4
Bank 3	6	2	0	8
Total	9	2	5	

Table 2.43 Bank balances at step 2.

	Equity	IBd	IBc	Net
Bank 1	2	4	9	7
Bank 2	2	4	2	0
Bank 3	2	8	5	−1

But with this new reduced value, bank 3 also defaults. With the same mechanism of residual wealth sharing among creditors, the new values will be as given in Tables 2.44 and 2.45.

What is important to note at this step is that we have feedback effects on bank 2, already defaulted, as a consequence of bank 3 defaulting. So the consequences of subsequent defaults propagate back not only to the safe banks but also to the defaulted banks, as shown in Tables 2.46–2.55.

Table 2.44 Interbank exposures at step 3.

	Bank 1	Bank 2	Bank 3	Total
Bank 1	0	0	4	4
Bank 2	3	0	1	4
Bank 3	5.25	1.75	0	7
Total	8.25	1.75	5	

Table 2.45 Bank balances at step 3.

	Equity	IB*d*	IB*c*	Net
Bank 1	2	4	8.25	6.25
Bank 2	2	4	1.75	−0.25
Bank 3	2	7	5	0

Table 2.46 Interbank exposures at step 4.

	Bank 1	Bank 2	Bank 3	Total
Bank 1	0	0	4	4
Bank 2	2.8125	0	0.9375	3.75
Bank 3	5.25	1.75	0	7
Total	8.0625	1.75	4.9375	

Table 2.47 Bank balances at step 4.

	Equity	IB*d*	IB*c*	Net
Bank 1	2	4	8.0625	6.0625
Bank 2	2	3.75	1.75	0.00
Bank 3	2	7	4.9375	−0.06

Table 2.48 Interbank exposures at step 5.

	Bank 1	Bank 2	Bank 3	Total
Bank 1	0	0	4	4
Bank 2	2.8125	0	0.9375	3.75
Bank 3	5.203125	1.734375	0	6.9375
Total	8.015625	1.734375	4.9375	

Table 2.49 Bank balances at step 5.

	Equity	IBd	IBc	Net
Bank 1	2	4	8.015625	6.015625
Bank 2	2	3.75	1.734375	−0.02
Bank 3	2	6.9375	4.9375	0.00

Table 2.50 Interbank exposures at step 6.

	Bank 1	Bank 2	Bank 3	Total
Bank 1	0	0	4	4
Bank 2	2.80078125	0	0.93359375	3.734375
Bank 3	5.203125	1.734375	0	6.9375
Total	8.00390625	1.734375	4.93359375	

Table 2.51 Bank balances at step 6.

	Equity	IBd	IBc	Net
Bank 1	2	4	8.00390625	6.00390625
Bank 2	2	3.734375	1.734375	0.00
Bank 3	2	6.9375	4.93359375	0.00

Table 2.52 Interbank exposures at step 7.

	Bank 1	Bank 2	Bank 3	Total
Bank 1	0	0	4	4
Bank 2	2.80078125	0	0.93359375	3.734375
Bank 3	5.200195313	1.733398438	0	6.93359375
Total	8.000976563	1.733398438	4.93359375	

Table 2.53 Bank balances at step 7.

	Equity	IBd	IBc	Net
Bank 1	2	4	8.000976563	6.000976563
Bank 2	2	3.734375	1.733398438	0.00
Bank 3	2	6.93359375	4.93359375	0.00

Table 2.54 Interbank exposures at step 8.

	Bank 1	Bank 2	Bank 3	Total
Bank 1	0	0	4	4
Bank 2	2.800048828	0	0.933349609	3.733398438
Bank 3	5.200195313	1.733398438	0	6.93359375
Total	8.000244141	1.733398438	4.933349609	

Table 2.55 Bank balances at step 8.

	Equity	IBd	IBc	Net
Bank 1	2	4	8.000244141	6.000244141
Bank 2	2	3.733398438	1.733398438	0.00
Bank 3	2	6.93359375	4.933349609	0.00

The Eisemberg and Noe algorithm has a correct estimation of the second and subsequent rounds of contagion, developing the computation of feedback effects at each step of contagion. This can be the best way for endogenizing the LGD, although the studies that used this approach are based on a fixed LGD.

With reference to the example given above, we saw that the losses to be shared are computed not only on the basis of the exposure, but also considering the default dimension. This allows the actual loss to be computed, and the initial exposure to be determined, the actual loss given default.

So the loss for bank 1 given default of bank 2 can be computed as the difference between the initial exposure, 6, and the final available value, 2.8, thus summing up to 3.2. The resulting LGD rate is 3.2/6, near 53.3%.

What changes with respect to the previous modeling is that, in this case, after bank 3 defaults at round 2, the higher losses for bank 2 will increase the losses for banks 1, 3, and 4. And this will increase the losses for bank 3 and so on up to convergence.

At the third round, when bank 4 defaults, the effects are to be computed also on bank 2, which will affect banks 1, 3, and 4, and for bank 3, this will affect banks 1, 2, and 4, and so on.

So this approach will include a recursive iteration that needs more time to be completed, but in the end it is possible to evaluate what the losses were found to be for each bank, so one can estimate the actual LGD for the simulated process.

The second approach, known as the sequential default algorithm, developed by Furfine (2003), was used much more often in contagion studies, possibly because of its simpler mechanism.

It involves the following steps:

1) One or more banks fail.
2) Any other bank fails by contagion if its exposure versus the defaulted bank(s), multiplied by an exogenously given LGD, exceeds its equity.
3) A second round of contagion occurs if there is any other bank for which the losses, due to interbank loans with defaulted counterparts, exceed its equity. Contagion stops when no additional banks go bankrupt. Otherwise, another round of contagion takes place.

Different from the Eisenberg and Noe algorithm, the Furfine algorithm (Table 2.56) does not include feedback effects for the banks that have failed previously. In fact, it is of no use to compute the precise value of the loss estimated by the contagion mechanism, if the actual loss is computed as the exposure multiplied by the (exogenous) LGD. So the papers assuming a constant or exogenous LGD often preferred this second sequential algorithm.

Table 2.56 Furfine sequential algorithm.

	Bank 1	Bank 2	Bank 3	Bank 4	Defaults	IB losses
IBc share	40%	20%	15%	25%		
Capital 1	10	−1	1	3	Bank 2	8
Losses 1	4		1.5	2.5		
Capital 2	6	−1	−0.5	0.5	Bank 3	5
Losses 2	2.4			1.5		
Capital 3	3.6	−1	−0.5	−1.0	Bank 4	3
Losses 3	1.6					
Capital 4	2.0	−1	−0.5	−1.0		

Acknowledging the limited information on the actual levels of LGD, usually papers tested the effects of contagion with different levels of this variable.

2.13 Data and Applied Problems

Data Sources
When dealing with simulations, as always in quantitative analyses, the quality of input data is fundamental. In this paragraph we will present some important issues referring to data and applied problems, in order to provide some hints and suggestions on how to operationally apply the methodologies described above.

The first problem refers to where to find the input data.

Even if it is possible to have it directly from the bank's balance sheet, it is simpler and more effective to download it from the banking-specific databases, as well as to have homogeneity in the standards and reproducibility in estimates.

One of the most commonly used data sets is Bankscope, a comprehensive global database of banks' financial statements by Bureau Van Dijk, also including a wide range of other banking information, such as ratings, ownership, and so on. Even if

Bankscope is a commercial database, and thus subject to access fees, many universities have access to it for research purposes and quite often extended to their graduate students.

As in the United States, where the FFIEC Uniform Bank Performance Report is publicly available, in many countries other free access databases are available, provided by either central banks or supervision institutions.

Bankscope includes a number of very useful different classifications. With reference to the banks included in the sample, apart from the home country, it is possible to select from different kinds of banking activities. In fact, central banks, investment banks, and bank holdings are also included, while the reference bank activity is mainly held by commercial banks, savings banks, and, depending on the considered country and banking system, other categories. So, the primary focus is on which banks to include in the analysis, and this must correspond with the specific aims of the analysis.

There must also be a focus on the interbank exposures. As even central banks are banks, even if their activity is evidently different, in some cases the interbank exposures include the exposures to the central banks. Here also, depending on the analysis-specific aim, it can be useful to include it or not, or to have a specific evaluation of the exposures to the central bank. It is in any case fundamental to verify if, for the considered country, the exposures with respect to the central bank are included, and this can be possibly done cross-checking against banks' balance sheets or with central banks' or international institutions' data.

Capital is another value that must be carefully considered, as more variables are used for measuring it, as equity, common equity, Core Tier1, Tier1, regulatory capital, total capital, and so on. In addition, some analyses focus on default, such as when the available capital is fully wiped out by the crisis, while others refer to distress, such as when the bank capital is reduced under the minimum capital requirement for acting with full activity. Here also it is important to select the capital definition corresponding more with the specific aims of the analysis.

Specific attention must focus on the evaluation of consolidated and unconsolidated statements.

For many banks, both consolidated and unconsolidated financial statements are available. This happens when the holding is itself acting as a bank, so it reports the holding bank activity as if it were a single bank, but also the consolidated statement that includes the activity of all the banks in the group. Quite often, depending on the single country's legislation, even the other subsidiary banks in the group publish the financial statement, and caution is advised regarding this issue for a number of reasons.

First, including in the sample all the available statements often results in double counting for some subsidiary or holding banks. To avoid this, it would be better, when both unconsolidated and consolidated statements are available, to first keep only the consolidated one, and then, if possible, verify which banks are parts of a group, thus already considered in a consolidated statement, and exclude those subsidiaries.

This is also important as quite often in banking groups, there is a diversification of roles, a centralized treasury, and a concentration of capital within the holding. So, some subsidiary banks can become undercapitalized, be hugely exposed to the interbank market, or carry high risks, but this only depends on the specific role on funding raising, or on treasury management, or on the lending strategy the bank plays within the group, which keeps the equilibrium in its whole organization, while considering single components sounds as unbalanced.

Another issue related to groups is that the within-group exposures are netted in the consolidated statements. This means that the consolidated statement only keeps the interbank exposures of banks in the group to banks outside of the group. The effect of considering, instead, also the exposures within the group as relevant for interbank contagion is evident.

Banks and Banking Groups

Large banking groups tend to differentiate their activity within the group, so that each bank plays a different role. The structures of the different groups are not standardized. In some cases, there is a geographical differentiation, so in some internationally active groups there is one bank working or coordinating the group activity in the country; in other cases, each business line is held by one bank of the group, so there is one reference bank for retail banking and another coordinates the investment activity or trading. Also the ownership can be different; in some cases the group head is the owner of the subsidiaries, while in other cases the local banks are the shareholders of the central holding.

In any case, within groups, there are channels for balancing profitability and quite often a centralization of liquidity management.

The profitability balancing within the group, thus the risk coverage by the group, makes the analysis of the risk affecting a bank part of a group biased when the group role is not considered, as also the capitalization can be concentrated into the holding (or in different ways), which allows a risk covering by the group, but does not allow the correct measure of the single bank capitalization. Some countries also do not allow the financial statement (balance sheet and income statement) of subsidiary banks within a group to be published if they are guaranteed by the group.

The same happens for liquidity. Banking groups tend to structure the liquidity management as in a money center structure (see Section 2.9), so as to simplify the liquidity interactions between subsidiaries. The actual interbank exposures of the group can be evaluated by means of the consolidated balance sheet, where the internal positions are netted, so the interbank deposits and loans only refer to the external counterparts of the group.

Considering a four-banks group, we can represent the interbank exposures, as in Table 2.57, in a matrix where lines refer to the creditor banks and columns to the debtor banks, plus the exposures outside of the group.

Considering the banks in the group as singles, we will find (Table 2.58), for example, that bank 1 has interbank credits for 180 and interbank debts for 650.

In the example, bank 1 register debts to bank 2 for 100, to bank 3 for 200, to bank 4 for 50, and to counterparts out of the group for 300. In bank 1's balance sheet, we will see that bank 1 has interbank credits for 180 and interbank debts for 650.

Table 2.57 Interbank group exposures.

	Bank 1	Bank 2	Bank 3	Bank 4	Out
Bank 1		10	50	20	100
Bank 2	100				
Bank 3	200	50		5	
Bank 4	50	10			
Out	300				

In this case, the actual relevance of the possible contagion channel is only for the "external" exposure, as in case of difficulties suffered by one bank in the group, the group itself will cover for it and the problems either have a serious impact on the whole group or are balanced by the group. Thus, accounting for the banks in the group as single banks, the contagion channels will be referred to the total exposures, including the intragroup liquidity management, resulting in an evident over-estimation of the actual group exposure.

System Definition

One of the important issues with reference to simulations is that when the interlinkages are considered, results are system dependent. It is thus important to have a clear consciousness of the

Table 2.58 Banks' balance sheet resulting exposures.

	Interbank credits	Interbank debts
Bank 1	180	650
Bank 2	100	70
Bank 3	255	50
Bank 4	60	25
Group (banks 1–4) total	595	795
Group consolidated	*300*	*100*

considered sample characteristics. The most commonly used definition of banking system refers to a single country. In this way, not only is the legislative and reporting framework homogeneous but also the hypotheses used for estimating the interbank matrix of exposures have a higher probability to only introduce an acceptable distortion. In fact, not only do the interbank exposures, in particular for small banks, tend to be held to counterparts of the same country but also the correlation (to macrovariables or among the considered banks) is higher and more coherently modeled when referred to only one country. Something similar happens when considering large countries (e.g., the United States) or economic areas (European Union), where large banks and banking groups deal with each other as in the higher level of the "multiple money center" structure (see Section 2.9). In any case, it is important to keep coherence with reference to the system definition, interbank channels and estimation, and macrovariables or shock induction, to yield a realistic representation and thus more realistic results.

Random Sequences

The Monte Carlo simulation requires a high number of iterations to yield a good approximation in its output probability distribution. But when dealing with tail risk, since the attention is not on the whole distribution but only on a small part, either a very large number of simulations or the use of a variance reduction technique is fundamental to yield acceptably precise results.

With regard to this aim, one technical detail is important. The random number sequence generated by the software typically comes from the iteration of a pseudorandom generator function, which always generates the same number sequence. This helps in comparing different models or settings, with the caution that in order to prevent any influence of the random generation, the numbers must be used by the software in the same sequence.

So, if in one model referring to five banks, we have five numbers generated for each simulation, and in another model

Table 2.59 Random numbers use: first case.

	Idiosyncratic factor 1	Idiosyncratic factor 2	Idiosyncratic factor 3	Idiosyncratic factor 4	Idiosyncratic factor 5
Iteration 1	0.662	−0.488	−0.311	−1.376	−1.694
Iteration 2	0.142	0.038	−0.554	1.345	0.860
Iteration 3	−0.126	0.160	−0.410	−0.732	−0.357
Iteration 4	0.267	−0.861	0.001	−1.913	−0.501

we add a macrovariable so that the software uses six numbers for each simulation, in practice the sequences are different.

If you need to have the same sequence, you can simply include in the software the generation of six variables in both cases, even if in the first case you only use five.

One example might help to explain the effect: In Table 2.59, we have the first case, five random values, while in Table 2.60 we have the second case. Even if the sequences of random numbers are the same (0.142 comes after −1.694), only the first simulation (iteration 1) is based on the same random numbers, while the others, due to the shifting of one number at each line, are computed on the basis of different inputs. If, instead, we use the second sequence for both simulations, just not considering the "common factor" values for the first case, we can neutralize the difference in the random number generation, which can be really useful for testing and debugging.

If, instead, the aim is to have the largest number of iterations, it is possible to "randomize" the sequence, so as to have a different

Table 2.60 Random numbers use: second case.

	Idiosyncratic factor 1	Idiosyncratic factor 2	Idiosyncratic factor 3	Idiosyncratic factor 4	Idiosyncratic factor 5	Common factor
Iteration 1	0.662	−0.488	−0.311	−1.376	−1.694	0.142
Iteration 2	0.038	−0.554	1.345	0.860	−0.126	0.160
Iteration 3	−0.410	−0.732	−0.357	0.267	−0.861	0.001
Iteration 4	−1.913	−0.501	0.839	−0.925	0.332	0.480

sequence for each set of simulations. This can be very useful for verifying the variability in results due to the random number generation, or to split the simulation process into smaller sequences, which can then be brought together at the end of the process, for example, first generating 10 sets of 1 million iterations each, possibly on different computers, for obtaining 10 million iteration sequences.

Simulations Results and Probability Distributions

The output of a Monte Carlo simulation is a vector containing the losses (or other values) computed for each simulation performed. The main part of the simulation (hopefully!) results in zero defaults. Therefore, normally it is not important to record these iteration results, but it is just important to know how many iterations resulted in zero defaults, to provide the correct picture of the probability distribution.

After this, there are a number of "small" defaults that do not start contagion. Then, an iteration starts contagion, and the last part of the distribution typically only includes contagious simulated crises. If the aim is to isolate the effect of contagion, it is possible to obtain it by running contagion and no contagion simulations before reordering, thus keeping the same input shocks for the same line of iterations, computing the difference between the two vectors.

3

Real Economy, Sovereign Risk, and Banking Systems Linkages

International connections between banks, and between the banking systems, public finances, and the real economy, are one of the important issues of the recent financial crisis.

The interlinkages between these three actors showed important effects not only on the direct connections but also on cross-effects and side effects, due to the multiple channels connections and the sensitivity to the same macroeconomic variables that induces correlation between them.

Considering the most evident contagion effects, public finances can be hit by banking crises as governments are expected to intervene with state support to prevent any spillover effects to both the banking system and the real economy. In turn, the weakening of public finances may affect the banking risk: The higher deficits (or even the contingent liability) typically induce a reduction in value of the sovereign bonds held by banks, which can possibly cause more bank stresses.

Moreover, the interconnections typically cross borders, so that a banking crisis in one country can induce state support and interbank freezing in some other country, or a haircut in one country's sovereign bonds can affect banks in different countries.

Another contagion channel is due to the sovereign bonds CDS market, these derivatives being issued and held mainly by banks, written on some country sovereign bonds, and guaranteed by other countries' sovereign bonds.

Vuillemey and Peltonen (2013) focused on the interplay between banks' bond and CDS holdings, by means of a

Banking Systems Simulation: Theory, Practice, and Application of Modeling Shocks, Losses, and Contagion, First Edition. Stefano Zedda.
© 2017 John Wiley & Sons, Ltd. Published 2017 by John Wiley & Sons, Ltd.

simulation model, and verified that in case of a sovereign credit event, banks' losses due to CDS exposures are less important than those due to bond exposures, while for CDS sellers, the main risk is due to the possible sudden increases in collateral requirements.

In a time of crisis, these effects are evidently increased by the "vertical" correlation of both the banking system and public finances to the real economy weakening, and the "horizontal" correlation between public finances and/or banking systems of different countries.

Considering these cross-linkages, quantifying the potential effects of interbank crises, and the effects that banking crises have on sovereign risk as well as the effects that a sovereign crisis has on the banking sector, is of fundamental importance in order to address the issue and restore financial stability.

While the direct transmission mechanisms have long been recognized, the existing literature, to date, has paid little attention to a quantitative analysis of the systemic effects considering the banking systems and public finances of different countries as a whole.

As defined by the Basel Committee on Banking Supervision in 2010, robust financial systems are those that do not adversely induce the propagation and amplification of disturbances affecting their soundness. As a consequence, a key issue for financial supervision and macroprudential regulation is understanding and quantifying the links between financial and public sectors.

An important reference for describing the contagion channels is the IMF (2010) representation (set out in Figure 1.5) of the interconnections between banks and sovereigns in a domestic and international perspective, even if it has not yet been translated into a formal model, and tested in some way.

3.1 Effects of Bank Riskiness on Sovereign Risk

The positive relationship between public debt and interest rates has been verified in many studies, as in Edwards (1986), Alexander and Anker (1997), Lemmen and Goodhart (1999),

Lønning (2000), Copeland and Jones (2010), and Codogno *et al.* (2003). Moreover, in the eurozone, sovereign spreads are recognized as being mainly driven by debt, deficits, and debt-service ratios (see Bernoth and von Hagen, 2004). In particular, Bernoth and Erdogan (2010) studied the determinants of sovereign bond yield spreads in 10 EMU countries between 1999 and 2010, including the recent financial crisis.

Linkages between bank distress and costs for public finances have been explored in terms of indirect effects via the real economy, impacts on national accounts due to state aid, and financial market reactions.

With reference to the indirect effects of bank distress via the real economy, Eschenbach and Shuknect (2002) analyzed the fiscal costs of financial instability, and Furceri and Zdzienicka (2010) looked at several episodes of banking crises in 154 countries over the period 1980–2006 in order to quantify the development of the gross government debt-to-GDP ratio in the aftermath of the banking crises. Among more recent papers, Reinhart and Rogoff (2008a, 2008b) gave an overview of the history of financial crises from the mid-fourteenth century to the 2008 subprime crisis. In a subsequent paper of 2009, the same authors focused on a comparison of the 2007 US crisis with previous episodes in terms of public debt. Laeven and Valencia (2010) estimated the costs of the 2007–2009 systemic banking crisis using three metrics: direct fiscal costs (fiscal outlays committed to the financial sector from the start of the crisis up to end of 2009), output losses (deviations of actual GDP from its trend), and increases in public sector debt relative to GDP (changes in the public debt-to-GDP ratio over the 4-year period beginning with the crisis year).

The direct influence of bank riskiness on public finances is mainly represented by the State support of the banking sector in case of financial crises. State aid often occurs through an injection of funds to a distressed banking sector. In the period between October 1, 2008 and October 1, 2011, the Commission approved aid to the financial sector for an overall amount of EUR 4.5 trillion (36.7% of EU GDP). Details of the actions taken by the European Union and the Member States to tackle the

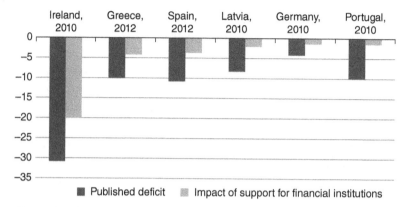

Figure 3.1 Increase in government deficits due to support for financial institutions (% GDP). *Source:* Reproduced with permission of Eurostat, Statistics in focus n. 10/2013.

financial and economic crisis can be found in the database available on the website of the European Commission Director-ate-General for Competition.[1] The largest impacts experienced by Member States are set out in Figure 3.1, where the support for financial institutions is expressed in terms of its impact on public deficit.

In European Economy (2011), an *ex ante* perspective is used for evaluating the risk of banking crises on national accounts (public deficit). Three indicators of sustainability are proposed and tested for Germany, Ireland, Portugal, and Sweden: the probability that public finances are hit by losses deriving from bank defaults, the distribution of costs for public finances, and the probability that, due to a banking crisis hitting public finances, a Member State becomes a high risk in terms of sustainability.

Recent empirical literature has looked at financial markets to assess the relationship of the banking risk to sovereign credit risk. Financial markets quickly internalize the effects described above: as observed in Acharya *et al.* (2011), as soon as financial sector bailouts were announced in the period 2007–2011, they were associated with an unprecedented widening of sovereign

---

1 See http://ec.europa.eu/competition/elojade/isef/index.cfm?
clear=1&policy_area_id=3.

CDS spreads and narrowing of bank CDS spreads. Thus, the contagion between the two sectors through the CDS market is very quick, different from the translation of banking crises into public debt via higher deficit or via the real economy where contraction of economic activity is mainly due to reduced lending and augmented funding costs (see Cecchetti *et al.*, 2009).

3.2 Effects of Sovereign Risk on Bank Riskiness

Sovereign risk refers to the risk that a government may default on its debt obligations. In the current European context, the term sovereign risk has been used to broadly categorize the large budget deficits and very high government debt levels of a number of countries, especially Greece, Italy, Ireland, Portugal, and Spain. An increase in a country's sovereign risk implies changes in sovereign bond yield spreads, sovereign credit default swaps spreads, or sovereign ratings.

The transmission of sovereign risk to banks' riskiness and profitability was clear in recent years, as the increase in sovereign risk and the downgrade of several countries had a negative effect on bank riskiness.

The direct impact of the recent sovereign debt crisis on banks' balance sheets was quantified in the European Banking Authority (EBA)'s 2011 technical proposal to the European Council. In its capital exercise, EBA was able to provide an overview of the sovereign portfolio of European banks, which represent the main channel of transmission.

As explained, when banks are creditors of the governments, the impact of a sovereign crisis on the banking system is often accentuated by the fact that government debt had been taken up increasingly by domestic banks in the run-up to a debt crisis. Once banking problems emerge, it is harder for governments to implement measures to contain a crisis given the reduction of both the available sources of funding and credibility.

With reference to the interlinkages between bank and public finances, Galliani and Zedda (2015) analyzed the cross-relationship and the possible crisis worsening due to the circular nature of the relationship. Duffie and Singleton (1999) and Ang and Longstaff (2013) use the sovereign credit risk channel to propagate sovereign-specific credit shocks and show that it induced a cascade of defaults in U.S. and Eurozone sovereigns.

Longstaff (2010) and Garratt *et al.* (2014) studied the relationship between reduced collateral values and asset price contagion. They showed that defaults in the subprime market quickly spread through the financial system and found that the role of asset-backed securities can be fundamental in the transmission of financial crises.

3.3 Linkages to the Real Economy

Regarding the linkages of the banking results to the real economy, while many studies have tested the influence of the banking activity on GDP, only a few refer to how the GDP variations influence the banking results, loans riskiness, and losses.

Karimzadeh *et al.* (2013) found that GDP has a positive relationship with ROA; they argued that the economic growth induces a higher demand for credit and vice versa.

Demirguc-Kunt and Huizinga (2000) and Bikker and Hu (2002) found that bank profits are correlated with the business cycle.

Athanasoglou *et al.* (2008) confirmed that the business cycle significantly affects bank profits, even after controlling for the effect of other determinants strongly correlated with the cycle (e.g., provisions for loan losses). They also tested for asymmetry in the effect of the business cycle, finding that the coefficient of cyclical output almost doubles when output exceeds its trend value, while it is insignificant when the output is below its trend.

Dietrich and Wanzenried (2011) in their study on the bank performances in Switzerland found that the business cycle significantly affects bank profits, so bank profits seem to be procyclical.

Albertazzi and Gambacorta (2009) specified that "bank profits pro-cyclicality derives from the effect that the economic cycle exerts on net interest income (via lending activity) and loan loss provisions (via credit portfolio quality)."

On the other hand, the effects of GDP variation on public finances are well known. Mourre *et al.* (2013) provided a good survey of the literature and the reference methods used by the European Commission for estimating the impact of a GDP variation on public expenditure and revenue.

Paltalidis *et al.* (2015) simulated negative shocks emerged from three possible sources: sovereign bonds, interbank lending, and customer loans. The initial shocks are induced as a 10% reduction in value of the considered assets category, and so the impact on each bank depends on its exposure to it. But the model also considers the subsequent effects, due to the fire selling of other assets for restoring at least the liquidity needed to meet the interbank liabilities. As this fire selling induces losses for the seller bank and value reductions for the asset category, the shock propagates through the system even when banks are not suffering direct losses in the interbank market.

In this way, contagion occurs either when a bank is insolvent (credit risk), thus for direct loss, or due to the fire sales value effect (market risk) when a bank needs to sell its sovereign bonds or loans at a lower price than expected. Therefore, for each of the induced shocks, contagion depends on the size of the bank, the distribution of exposures, the degree of interconnectedness, and the capitalization level.

This modeling highlights that as the bank balance sheet must meet a number of minimal equilibriums, any distress episode can possibly transfer risks and losses from one asset category to the others, even when the categories are not directly linked. In fact, even if the markets are distinct for sovereign bonds, customer loans, and interbank lending, as the actors are often the same, they become connected through balance sheet contagion. The same approach can be extended to include CDS and other derivatives, or even the liabilities side, and possibly deposits.

Huang *et al.* (2009) examined the dynamic linkages between default risk factors and a number of macrofinancial factors. They designed an integrated micro–macro model framework to examine the determinants of PDs and correlations, thus enabling the two-way linkages between the banking sector and the macrovariables to be investigated. The model estimation consists of two equations.

The first equation, referring to the macro side, adopts a VAR framework that allows for dynamic linkages between the credit risk factors of the banking system and a list of macrofinancial variables related to the evolution of the macroeconomy and of financial markets.

The second equation refers to the micro side and models the default risk of individual banks by the credit risk factors of the financial system and financial market variables, thus explaining the movements of individual PDs as the response to changes in market conditions.

$$X_t = c_1 + \sum_i b_i X_{t-i} + \epsilon_t \tag{3.1}$$

$$\text{PD}_{i,t} = c_{2i} + a_i \text{PD}_{i,t-1} + \gamma X_t + \mu_{it} \tag{3.2}$$

where the X variables include the credit risk factors (average PD and 1-week correlations) in the banking sector and macrofinancial variables. The results, combined with the estimated correlations, determine the economic system dynamics.

3.4 Modeling

3.4.1 Banks

Bank i's balance sheet can be represented as follows:

Assets		Liabilities and equity	
Loans $\sum_k A_{ik}$		Equity K_i	
Sovereign bonds $\sum_c \text{SB}_{ic}$		Deposits	
Interbank credits $\sum_j \text{IB}_{ji}$		Interbank debts $\sum_j \text{IB}_{ij}$	
Other assets		Other liabilities	

Here, c refers to countries, k to customer loans, and j to the counterparty banks.

As in the Basel II FIRB model, bank results can be represented as a one-factor model, partially determined by diversified risks, thus determined by the assets PD, but partially dependent on a nondiversifiable component.

The diversified component is quantified as the losses expected value, given by $\sum_k A_{ik} \times PD_{ik} \times LGD_{ik}$, where, for each loan k, A_{ik} is the amount of the loan, PD_{ik} is its probability to default, and LGD_{ik} is the loss given default.

The undiversified component, linked to the macro variables, can be proxied by the GDP variation of the home country. As banks are typically active in more countries, the reference variable can be better represented as the weighted average of the GDP variations of the countries where the bank is investing.

So, for the bank i active in the countries c with shares of its activities D_{ic}, the loan losses L_i can be represented as follows:

$$L_i = f\left(\sum_k A_{ik} \times PD_{ik} \times LGD_{ik}, \sum_c D_{ic} \times GDP_c, \epsilon \right)$$

$$(3.3)$$

where D_{ic} is the country weight for the bank i, so that

$$\sum_c D_{ic} = 1 \qquad (3.4)$$

and ϵ represents the idiosyncratic component of bank results.

Banks are considered to be in distress as soon as losses become higher than capital:

$$L_i > K_i \qquad (3.5)$$

The second source of losses is the interbank contagion, so that in case of distress of bank j, bank i, creditor for the amount IB_{ji}, will experience a loss given by $IB_{ji} \times LGD_j$, where LGD_j is the interbank loss given default (LGD) of the distressed bank j.

Considering all distressed banks, we have, for bank i:

$$L'_i = L_i + \sum_j \left[d(j) \times \text{IB}_{ji} \times \text{LGD}_j \right] \tag{3.6}$$

where $d(j)$ is set to 1 for the distressed banks and 0 for the nondistressed banks.

The third component of losses is due to the sovereign bonds market value reduction (haircut).

3.4.2 Public Finances

In the literature, the value of sovereign bonds is mainly referred to three components, namely, the country's size, the country's debt, and the country's deficit. While the first two are rather stable in a 1-year horizon, the third is affected both by the GDP variations and by the possible interventions for bank rescuing.

The deficit variation is the result of the projected public budget result, $\overline{\text{DEF}_c}$, corrected for the effect of the GDP variation:

$$\text{DEF}_c = \overline{\text{DEF}_c} + \Delta\text{DEF}_c(\Delta\text{GDP}_c) \tag{3.7}$$

This can be proxied as the GDP variation multiplied by the sensitivity parameter of deficit to GDP for the country, S_c:

$$\text{DEF}_c = \overline{\text{DEF}_c} + \Delta\text{GDP}_c \times S_c \tag{3.8}$$

and

$$\Delta\text{DEF}_c = \Delta\text{GDP}_c \times S_c \tag{3.9}$$

The sovereign bonds will then be affected by a variation in its market value, ΔSB_c, from the original value SB'_c to the new value SB_c as

$$\Delta\text{SB}_c(\Delta\text{DEF}_c) = \text{SB}'_c - \text{SB}_c \tag{3.10}$$

As a consequence, every bank i that has invested in sovereign bonds of the country c will experience an additional loss

$$\Delta\text{SB}_{ic}(\Delta\text{DEF}_c) = \text{SB}'_{ic} - \text{SB}_{ic} \tag{3.11}$$

and summing up for all countries we have

$$L_i'' = L_i + \sum_j \left[\mathrm{IB}_{ji} \times d(j) \times \mathrm{LGD}_j \right] + \sum_c \Delta\mathrm{SB}_{ic} \qquad (3.12)$$

where $d(j)$ is the dummy variable set to 1 in case of bank j defaulting and 0 otherwise.

The second source of public finances instability we consider is the cost of bank rescuing.

Based on the previous paragraph representation, we have

$$\mathrm{DEF}_c' = \mathrm{DEF}_c + \sum_i d(i) \times \left(L_i'' - K_i \right) \qquad (3.13)$$

where $d(i) = \begin{cases} 1 & \text{if} \quad L_i'' > K_i \\ 0 & \text{if} \quad L_i'' \leq K_i \end{cases}$

This additional public deficit will itself affect the bank's stability, inducing more defaults, more contagion, more need to bank rescuing, and so on.

3.5 Implementation

3.5.1 Public Finances

The public finances deficit is first simulated as

$$\mathrm{DEF}_{cs} = \overline{\mathrm{DEF}_c} + \Delta\mathrm{DEF}_{cs}(\Delta\mathrm{GDP}_{cs}) \qquad (3.14)$$

where $\overline{\mathrm{DEF}_c}$ is the base reference value and $\Delta\mathrm{DEF}_{cs}(\Delta\mathrm{GDP}_{cs})$ is the simulated deficit variation for the country c in simulation s.

This can be proxied as the GDP variation multiplied by the sensitivity parameter of deficit to GDP for the country, S_c:

$$\mathrm{DEF}_{cs} = \overline{\mathrm{DEF}_c} + \Delta\mathrm{GDP}_{cs} \times S_c \qquad (3.15)$$

and

$$\Delta\mathrm{DEF}_{cs} = \Delta\mathrm{GDP}_{cs} \times S_c \qquad (3.16)$$

where S_c is the sensitivity of the country balance sheet to GDP variations, as estimated by the European Commission, DG ECFIN for surveillance purposes.[2]

$\Delta GDP_{cs} \sim N(\mu_c, \sigma_c^2)$ is a randomly generated variable with mean and variance calibrated as in the last 10 years for the considered country. The country GDP variation is correlated among countries, also for the simulated GDP variation, on the basis of the actual results for the last 10 years. Thus,

$$\text{corr}(\Delta GDP_{cs}, \Delta GDP_{gs}) = \gamma_{cg}$$

where γ_{cg} is the actual correlation of the GDP yearly variations between country c and country g observed in the last 10 years.

3.5.2 Banks

While some of the values are in the bank's balance sheet and can be easily obtained from databases like Bankscope, the distinction of sovereign bonds in each bank portfolio by the issuing country is typically not reported but for the main European banks it can be obtained from the EBA.[3]

One important issue refers to the bank's assets riskiness, crucial for modeling the loan losses probability distribution.

In the previous literature two references are considered: the first is based on the loan losses provisions, while the second is based on historical data on previously suffered losses.

In the absence of an effective intervention by resolution facilities, whenever a bank defaults, it is assumed that part of its interbank debts are passed as losses to creditor banks.

Recalling that

$$L_i' = L_i + \sum_j \left[IB_{ji} \times d(j) \times LGD_j \right] \tag{3.17}$$

where $d(j)$ is set to 1 for the distressed banks and 0 for the nondistressed, and LGD_j is the interbank LGD of the distressed

---

2 See Mourre *et al.* (2013).
3 See EBA capital exercise.

bank, we will estimate contagion based on

$$L'_{is} = L_{is} + \sum_j \left[\text{IB}_{ji} \times d(j_s) \times 0.4 \right] \qquad (3.18)$$

where

$$\text{IB}_{ji} = \sum_i \text{IB}_{ji} \times \frac{\sum_j \text{IB}_{ij}}{\sum_i \sum_j \text{IB}_{ij}} \qquad (3.19)$$

is the share of the interbank debts of the distressed bank *j* attributed to the *i* bank.

The sovereign bonds haircut is estimated as follows:

$$\Delta \text{SB}_c(\Delta \text{GDP}_c) = \Delta \text{GDP}_c \times S_c \times \text{YS} \times \text{YP}_c \qquad (3.20)$$

where S_c is the sensitivity of the country balance sheet to GDP variations, as estimated by the European Commission, DG ECFIN for surveillance purposes,[4] YS is the yield sensitivity to deficit variations,[5] and YP_c is the parameter for converting the yield variation into value variation preserving the market equilibrium of expected returns.

Adding the public finances component, the first round contagion will be

$$L''_{is} = L_{is} + \sum_j \left[\sum_i \text{IB}_{ji} \times \frac{\sum_j \text{IB}_{ij}}{\sum_i \sum_j \text{IB}_{ij}} \times d(j_s) \times 0.4 \right]$$

$$+ \sum_c \Delta \text{GDP}_{cs} \times S_c \times \text{YS} \times \text{YP}_c \qquad (3.21)$$

---

4 See Mourre *et al.* (2013).
5 See Bernoth and Erdogan (2010).

4

Applications

The actual use by regulation and supervision authorities comes from the importance of contagion effects spreading out from default or distress, coupled with the limited number of actual cases—so there is a lack of data. Simulations can solve this problem, as it is possible to produce as many cases as needed, and quite simply develop what-if tests on the effects of the variation of every considered variable. This can be really useful for the authorities to test new regulation or for tuning it.

It must be remarked that reliability of simulations results is based on the quality of the subjacent models, frameworks, and settings.

4.1   Testing for Banks–Public Finances Contagion Risk

One important issue coming from the recent financial crisis in Europe was related to the weakness of banks and public finances, also known as "the vicious circle" of banks and their sovereign. In fact, the banks' weakness was often backed by public finances, generating in the best cases a contingent liability, but in many cases an important recapitalization or direct financial support. The cost of this support increased the public deficit of the countries, already weakened by the first wave of the crisis. So, as the higher public deficit typically results in higher rates and lower values for the sovereign bonds, the effect came back to the banking sector, which typically owns sovereign bonds.

Banking Systems Simulation: Theory, Practice, and Application of Modeling Shocks, Losses, and Contagion, First Edition. Stefano Zedda.
© 2017 John Wiley & Sons, Ltd. Published 2017 by John Wiley & Sons, Ltd.

This circular mechanism, from banking weakness to public finances weakness and going back to banks weakness, seriously worried the European authorities, which intervened with new regulation for cutting this feedback loop and excluding the risk of a twin default of banks and sovereigns.

In this line, simulations were also used for assessing the impact on the main banks of possible haircuts on sovereigns (Zedda *et al.*, 2012b), performed on 65 large EU banking groups (Table 4.1).

This exercise, based on the estimation of the increase in banks' probability to default (PD), shows that the possible haircuts on sovereign debts of EU Member States would heavily worsen the stability of their banking systems, but could also have spillover effects on the financial stability of other EU countries (Table 4.2).

As expected, the major effect of the haircut is within the country. Important spillover effects can be observed in the case of a haircut on Greek sovereign debts for Cyprus banks, and Italian sovereign debts for Belgian and Luxembourg banks. The effect is clearly nonlinear in the amount of the haircut, due to the threshold effect exercised by the capital level of the bank owning the sovereigns.

On the other hand, a test for the effectiveness of the newly introduced bail-in tool (strong limits in the public interventions for bank rescuing) for breaking the vicious circle, based on a typical what-if exercise, comparing the old framework with the new one, is developed in Galliani and Zedda (2015) (Figure 4.1).

This second exercise shows that the risk of feedback loops was actually present in the old framework, but the bail-in tool introduction proves to be really effective in preventing the banks–public finances contagion, so that the feedback loop risks have disappeared, and banking crises are no longer affecting sovereign bonds values.

4.2 Banking Systems Regulation What-If Tests

In European Commission (2011a), in order to assess the stability of the public finances, the risk of possible banking crises on

Table 4.1 Banking group exposures to sovereign debts, by country as a percent of Tier 1 capital as of September 2011.

Country	Greece	Ireland	m€		
			Italy	Portugal	Spain
AT	0.5%	0.2%	4.4%	0.1%	0.4%
BE	13.2%	1.2%	57.2%	6.2%	8.1%
CY	87.3%	6.4%	1.8%	0.0%	0.0%
DE	4.2%	0.6%	19.6%	2.6%	12.5%
DK	0.2%	0.5%	4.2%	0.1%	1.9%
ES	0.2%	0.0%	5.2%	2.5%	124.4%
FI	0.0%	0.8%	0.0%	0.0%	0.0%
FR	3.7%	0.9%	20.6%	1.8%	5.6%
HU	0.0%	0.0%	0.0%	0.0%	0.0%
IE	0.2%	91.6%	1.8%	0.7%	0.2%
IT	1.5%	0.2%	160.5%	0.3%	3.9%
LU	5.9%	0.0%	100.0%	10.3%	12.4%
MT	2.0%	2.0%	1.7%	0.6%	0.0%
NL	1.0%	0.4%	5.0%	0.7%	1.5%
NO	0.0%	0.0%	0.0%	0.0%	0.0%
PL	0.0%	0.0%	0.0%	0.0%	0.0%
PT	4.9%	2.6%	4.6%	109.8%	0.5%
SE	0.1%	0.0%	0.2%	0.1%	0.0%
SI	0.4%	0.6%	1.8%	0.2%	1.1%
UK	0.5%	0.3%	7.8%	0.6%	2.5%
Total	2.4%	1.5%	24.4%	3.3%	18.3%

Source: Zedda, 2012. Reproduced with permission of European Union.

national accounts is evaluated from an *ex ante* perspective. In this case, the focus was in particular on the effect of the proposed new regulation that included higher capital requirements, the elimination of the government bailout of distressed

Table 4.2 Weighted average probability of default of the banking groups in each country under possible haircuts of some countries sovereign bonds.

	No haircut	Haircuts				
		GR 30%	IE 30%	IT 30%	PT 30%	ES 30%
AT	0.130%	0.131%	0.131%	0.137%	0.131%	0.131%
BE	0.091%	0.111%	0.092%	0.217%	0.100%	0.095%
CY	0.094%	0.315%	0.102%	0.097%	0.094%	0.094%
DE	0.099%	0.103%	0.100%	0.117%	0.101%	0.105%
DK	0.057%	0.057%	0.057%	0.060%	0.057%	0.057%
ES	0.169%	0.169%	0.169%	0.178%	0.174%	0.358%
FI	0.157%	0.157%	0.159%	0.157%	0.157%	0.157%
FR	0.155%	0.160%	0.156%	0.188%	0.157%	0.159%
HU	0.050%	0.050%	0.050%	0.050%	0.050%	0.050%
IE	0.292%	0.293%	0.708%	0.297%	0.294%	0.292%
IT	0.168%	0.170%	0.168%	2.942%	0.169%	0.171%
LU	0.011%	0.011%	0.011%	0.025%	0.011%	0.011%
MT	0.042%	0.043%	0.043%	0.043%	0.042%	0.042%
NL	0.102%	0.104%	0.103%	0.109%	0.104%	0.103%
NO	0.153%	0.153%	0.153%	0.153%	0.153%	0.153%
PL	0.192%	0.192%	0.192%	0.192%	0.192%	0.192%
PT	0.229%	0.247%	0.236%	0.238%	1.056%	0.230%
SE	0.183%	0.183%	0.183%	0.183%	0.183%	0.183%
SI	0.163%	0.163%	0.164%	0.166%	0.163%	0.164%
UK	0.092%	0.092%	0.092%	0.100%	0.092%	0.093%
Total	0.128%	0.132%	0.134%	0.349%	0.140%	0.147%

The average is weighted on total assets.
Source: Zedda, 2012. Reproduced with permission of European Union.

banks, the introduction of a resolution fund (RF) for limiting financial contagion, and other interventions. In order to assess the expected effects of these interventions in reducing the risk of public finances destabilization, the European Commission

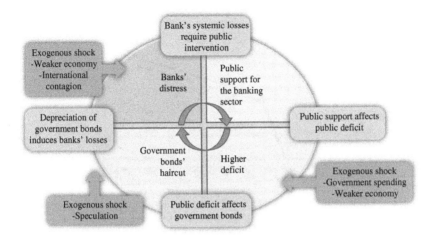

Figure 4.1 Banks and public finances interlinkages from a domestic perspective. *Source:* Galliani, 2015. Reproduced with permission of Springer.

developed a what-if exercise, based on simulations, with the aim of estimating the probability distributions of aggregate banking losses by country, and their expected impact on government finances. The estimation considered five scenarios (see Table 4.3), starting from the precrisis settings and following the proposed path for the gradual introduction of new rules, up to the full implementation of the new framework.

Table 4.3 Scenario definitions.

Scenario	Capital setting			DGS/RF setting		Bail in		Contagion	
	Basel II	Basel III 8%	Basel III 10.5%	Yes	No	Yes	No	Yes	No
1									
2									
3									
4									
5									

Source: Reproduced with permission of European Commission, 2011.

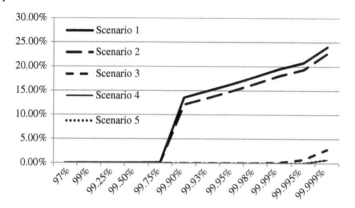

Figure 4.2 Estimated incidence on GDP of costs for public finances due to possible banking crises by probability level for Germany. *Source:* Reproduced with permission of European Commission, 2011.

For estimating the outcomes of the different scenarios in terms of public finances sustainability, the exercise considers three different indicators: the probability that losses deriving from bank defaults can hit public finances, the distribution of costs for public finances, and the probability that a European Union Member State becomes high risk in terms of sustainability due to a banking crisis. The method is tested on four countries: Germany, Ireland, Portugal, and Sweden.

Figure 4.2 and Table 4.4 report some of the results obtained in this exercise.

Table 4.4 Probability that one country becomes high risk in terms of public finances sustainability due to banking crises.

	DE	IE	PT	SE
Scenario 1	7.38%	2.64%	1.32%	0.41%
Scenario 2	0.11%	0.34%	0.27%	0.08%
Scenario 3	0.01%	0.13%	0.15%	0.04%
Scenario 4	0.00%	0.03%	0.06%	0.02%
Scenario 5	0.00%	0.01%	0.03%	0.02%

Source: Reproduced with permission of European Commission, 2011.

Apart from the values coming from these estimations, what is anyway important is the relevance for decision-makers of having this additional information for evaluating the expected effects of some important intervention.

The scenarios above report the different steps of introducing a new regulation framework, but are also a way for evaluating the differential effects of each step on the banking system riskiness, and on its possible effects on public finances stability. This is fundamental for evaluating the opportunity of actually set up the new regulation, even if a more effective analysis has to be based in cost-benefit terms, and evidently the final evaluation must consider all the effects of it, in a much broader impact analysis.

4.3 Banks' Minimum Capital Requirements: Cost–Benefit Analysis

After the 2008 financial crisis, a debate was opened on how to reduce the risk of a further worsening of the crisis, and on how to avoid, or at least reduce the risk of, another financial crisis.

One of the possible interventions that were immediately considered for reducing that risk was increasing the minimum capital requirement for banks.

In fact, a higher capitalization lowers the default risk of each bank, so the expected result of rising minimum capital requirements is of a higher stability of the whole system.

But this higher capitalization also induces effects on the credit conditions, as the cost of capital is typically higher than the cost of other funding sources, so the higher cost of funding is expected to induce higher interest rates spreads, thus higher costs for customer investments, and lower support to the economic growth.

So, the effects of higher capital requirements must be considered in cost–benefit balancing terms.

Marchesi *et al.* (2012) analyzed this issue, by means of a simulations-based model, whose approach and results are reported here, and obtained an interesting assessment of the expected results within the eurozone.

4.3.1 Costs

With reference to costs, the value of the higher capital needed can be obtained computing the difference from the actual capitalization level and the new minimum capital requirement for each bank.

The costs to the real economy of higher capital levels are more complex to estimate. In fact, this higher capitalization induces a higher cost for liability remuneration, as capital is the most expensive funding source.

In formal terms, we have to quantify the loan cash flow in and cash flow out, so as to find the loan pricing equilibrium point.

The credit risk related to the exposure implies that either the debtor defaults, with probability PD, or the debtor regularly pays the credit, with probability $1 - PD$.

Starting from the standard reference of a unitary credit, with 1 year maturity, and interests paid at the end of the period, the expected value of the exposure j is given by

$$E(L) = \left(1 + i_j\right) \times \left(1 - PD_j\right) + \left(1 + i_j\right) \times \left(1 - LGD_j\right) \times PD_j$$

$$(4.1)$$

Thus,

$$E(L) = \left(1 + i_j\right) \times \left(1 - LGD_j PD_j\right) \qquad (4.2)$$

where

$E(L)$ is the expected value of the credit at the end of the period;
i_j is the interest rate applied on the j risky–loan;
PD_j is the probability of default of the j debtor;
LGD_j is the loss given default of the j debtor.

On the other hand, the loan price should cover the cash flows resulting from the funding needs and operative costs (credit-worthiness analysis, monitoring, etc.) related to the credit exposure.

Formally, it can be described as

$$U(M) = C_j(1 + r_e) + \left(1 - C_j\right) + (1 + i_d) + cop_j \qquad (4.3)$$

and

$$U(M) = C_j(r_e - i_d) + 1 + i_d + \text{cop}_j \qquad (4.4)$$

where

$U(M)$ is the overall cash flows out;
C_j is the capital funding rate;
r_e is the gross return to shareholders;
i_d is the interest rate paid on interbank funding;
cop_j is the operative costs related to the loan.

By jointly considering (4.1) and (4.2), we can get the break-even price of the loan:

$$(1 + i_j) \times (1 - \text{LGD}_j \text{PD}_j) = C_j(r_e - i_d) + 1 + i_d + \text{cop}_j$$

$$(4.5)$$

As the target variable is the loan interest rate, we will have

$$i_j = \frac{i_d + \text{PD}_j \times \text{LGD}_j + C_j(r_e - i_d) + \text{cop}_j}{1 - \text{LGD}_j \text{PD}_j} \qquad (4.6)$$

So, any variation in the capital level will affect the interest rate spreads in proportion to the difference between the equity remuneration rate and the interbank loans rate $(r_e - i_d)$.

Therefore, a higher capital coverage will raise the interest rate spreads, inducing higher costs for investments, lower investing, and lower economic growth.

Marchesi *et al.* (2012) estimated the costs in terms of lower GDP for a sample of seven European countries by the following formula, which also includes the taxation effects:

$$\Delta C_j(r_e - i_d(1 - \text{tax})) \times \left[(1 - \text{tax}) \times \text{leverage} \right] \times \left(\frac{\epsilon_{\text{CoC}}^Y}{\text{CoC}} \right) \times \text{DF}_\infty$$

$$(4.7)$$

where

ϵ_{CoC}^Y is the GDP elasticity to the cost of capital;
CoC is the cost of capital;

DF_∞ is the discount factor for an infinite stream of annual amounts.

In this formula, the first term accounts for the capital value variation multiplied by the difference between the return on equity and the interbank interest rate, as seen above, corrected for the bank taxation effects. The term in square brackets accounts for the firm taxation and exposure effects, so it translates the higher lending spreads into firms' higher costs. The last ratio further translates the higher costs in its effects on GDP, and the discount factor just provides for the translation of annual costs into the net present value of the future effects (see Bank of England, 2010).

It is important to highlight that the effects on GDP must be estimated separately for each country, as the taxation and leverage are different among the considered countries, with taxation rates ranging from 7.6% (Ireland) to 34% (Spain) and leverage ranging from 31.5% (France) to 57.8% (Italy).

4.3.2 Benefits

With reference to benefits, these are based on the lower occurrence of new financial crises, thus on the quantification of the cost of a financial crisis and on the variation of the estimated occurrence of crises.

The methodology used for quantifying the cost of each crisis is based on the IMF (2009) event case studies of the previous systemic crises, comparing the medium-term level of output to the level it would have reached following the precrisis trend, with the medium term defined as 7 years after the crisis. The fundamental result of these evaluations is that "the path of output tends to be depressed substantially and persistently following banking crises, with no rebound on average to the precrisis trend over the medium term" (IMF, 2009).

So, the cost is due to the immediate impact of the GDP reduction plus the discounted effects (net present value) of the lower growth trend for the subsequent years.

While the estimations of the single bank risk reduction can be obtained from the probability distribution of bank losses and capital coverage, the systemic effects, and thus the inclusion of contagion risks, suggest using simulations as an actual tool for quantifying the benefits in terms of risk reduction.

In this exercise, a sample of banks from each of the considered seven EU Member States was used for simulating the system stability with different capitalization levels and quantifying the risk of new systemic crises with total losses higher than of 3% of the country GDP.

In technical terms, this is obtained for each country by quantifying the 3% GDP threshold for the country (rescaled on the sample), and then counting for the simulations that report a losses volume higher than the threshold.

For each of these crises, the estimation of the cost was based on the discounted value of the expected GDP reduction estimated for the last crisis.

The graph in Figure 4.3 reports the main results of the analysis, for MCR levels of 6, 8, 10.5, 12, 13.5, and 15%.

While, in the formula, the impact on costs is linear with increasing capital coverage, the higher capitalization requirement

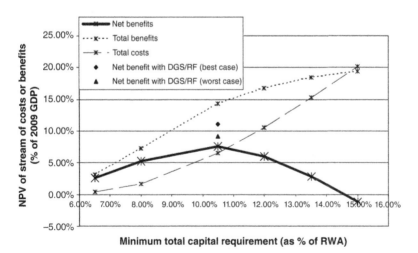

Figure 4.3 Costs, benefits, and net benefits of different levels of minimum capital requirements for banks. *Source:* Marchesi, 2012. Reproduced with permission of European Union.

impact appears to be lower for the leftmost part of the graph. This is due to the fact that in the cited analysis, the starting point is the actual capital coverage, and almost always banks have more capital than required, often already complying with higher requirements, so that only part of the higher capitalization must be actually raised from the market. As the MCR rate increases, costs reveal a linear relationship to the minimum capital requirements.

Instead, the impact on benefits, due to a lower crisis risk, is almost linear in the first part and then decreasing, due to the lower marginal impact of more capital coverage.

As a final result, the net benefits curve reports an optimal point near the 10.5% value, which was set as the basis of the new regulation.

4.4 Deposits Guarantee Schemes (DGS)/Resolution Funds Dimensioning

4.4.1 DGS

As already outlined in Section 1.1, bank runs are one of the major risks for banking systems stability.

The risk of bank runs is related to the maturity transformation, as banks typically borrow money as demand deposits, and invest in illiquid loans, which are impossible to sell quickly without a loss in value. When depositors lose confidence in a bank's stability, the normal reaction is to withdraw their deposits as soon as possible. This causes an important need for liquidity, and in order to restore the liquidity equilibrium the banks have to sell off first the liquid assets and then the profitable assets, thus causing important losses and possibly causing the bank to default.

Bank runs can also start without any actual bank difficulty, for example, if a bank is suspected of running into difficulties, in particular during major financial crises, when the lack of confidence not only affects distressed banks but also affects the other banks. In fact, quite often, the bank run is the cause of the actual default, so in this case, the suspicion for a possible default causes the actual default, even when the initial cause was not due to bank difficulties, just suspicion.

If caution is not exercised with this problem, even the false suspicion that a bank is likely to fail can cause a bank run, bringing the bank to actually fail!

That is why almost all countries have implemented a Deposits Guarantee Scheme, so as to guarantee that even in the case of a bank default, deposits are returned to depositors. In this way, depositors know that their values are covered by a guarantee, and the rumors of a possible bank failing induce fewer worried and nervous reactions.

Design of DGSs varies across countries. The main differences are related to the funding mechanisms: some countries set up an *ex post* funding mechanism.

The *ex post* funding is a simpler structure, where after the bank default the other banks contribute to the need to refund the deposits, so this does not require a large structure and funds management during "normal" economic periods. Instead, during major crises, this structuring can have procyclical effects, as the cost of recovering deposits is passed to the other banks, possibly already weakened by the crisis, and inducing a kind of institutional contagion.

In order to avoid these procyclical effects, other countries prefer *ex ante* funding, based on the contributions raised during "normal" periods so as to be effective during crisis periods.

Apart from the mechanism, for effectively playing its role, the DGS must have adequate funding, in order to promptly intervene in case of bank difficulties, when the financial system tends to be unstable and the lack of confidence tends to spread.

The problem of adequately dimensioning DGSs is fundamental, as an undersized funding will not convey the confidence of depositors, which is the aim of the DGSs, whereas oversized funding is not only inefficient but also not appreciated by banks, as DGSs are funded by banks' contributions.

So, it is important to evaluate which is the optimal dimension of DGSs, and this must be based on the actual risks coming from the covered banks, in terms of the probability distribution of expected DGS interventions.

The first evaluation to be conducted for a correct dimensioning of DGSs is in determining, or at least approximating, the value of covered deposits. In fact, typically DGSs only guarantee a part of the total deposits volume, with limits on the counterpart, for example, no guarantee for public and government deposits, and on the value, for example, no coverage for deposits above US$100,000.

It is, thus, normally not possible to have a precise evaluation of the actual volume of covered deposits from banks' balance sheets. Instead, often some estimation by supervisors or national and international institutions on the share of covered deposits at least for the country is publicly available. More accurate estimations must be based on a specific evaluation on single customer accounts in each bank.

The second problem is in the estimation of the probability distribution of DGSs interventions. This can be developed by means of simulations, as it allows the frequency and size of crises to be determined. The only important difference when doing simulations for DGS dimensioning is that the crisis outcome must not be quantified in terms of bank losses, but the value to be considered is in its covered deposits. In fact, the DGS intervention is not in covering losses, but in providing for the liquidity of the covered deposits, to avoid financial panic and irrational behavior of depositors and markets, and the intervention must be prompt and visible to be reassuring.

In applied terms, the estimation of the DGS probability distribution of interventions can be obtained starting from the defaults matrix (as in Table 4.5, derived from Table 2.6) with the value of covered deposits of each defaulted bank, and then summing on rows. The resulting vector, reordered, gives the estimation of the envisaged probability distribution.

The probability distribution can be used for determining either the share of possible crises that can actually be covered with a given DGS funding level or the funding needs for covering a given share of possible crises (resulting in a value at risk (VaR) estimation). In the example in Table 4.5, for covering 80% of cases a minimum of 330 is required, while a fund of 200 will only cover for 60% of cases.

This approach has been developed by De Lisa et al. (2011) with reference to Italy, and applied by Campolongo et al. (2010) for some European countries (Table 4.6).

Table 4.5 DGS simulated intervention values.

	Bank 1	Bank 2	Bank 3	Bank 4	Bank 5	Total
Iteration 1	200					200
Iteration 2			250	20		270
Iteration 3						0
Iteration 4			250		80	330
Iteration 5						0
Iteration 6						0
Iteration 7	200	100	250	20		570
Iteration 8						0
Iteration 9				20		20
Iteration 10	200	100	250	20		570

The second problem related to DGSs is in how to quantify the contribution of each bank to the funding needs.

The simple approach that quantifies contributions on the basis of the volume of covered deposits can be a first approximation, but limited by the irrelevance of differences in the actual

Table 4.6 DIS conditional loss distribution for the United Kingdom, Germany, and Spain (m€).

Percentile	UK	DE	ES
25%	1,892	7,312	4,175
50%	22,334	15,936	8,559
75%	262,201	140,119	17,288
90%	1,115,327	1,119,432	42,927
95%	4,180,117	1,119,432	136,284
99%	4,180,117	1,119,432	1,350,573
Mean	453,814	290,382	57,703
Std. dev.	1,078,527	464,383	229,909

Source: Campolongo, 2010. Reproduced with permission of European Union.

riskiness of different banks. In fact, lowering assets riskiness or rising capitalization, *ceteris paribus*, reduces the risk of bank default, hence the risk of a DGS intervention. Thus, the actual bank riskiness should be considered when quantifying the DGS contributions. In fact, regulators often request or at least suggest that contributions should be based on the actual risk borne by each bank.

The quantification of risk-based contributions is thus the target, and this also can be obtained by means of simulations, just focusing on column totals in the simulated DGS intervention matrix, thus on the actual default probability of each bank, times the covered deposits volume.

4.4.2 Resolution Funds

A similar problem refers to the possibility of setting up a structure similar to a DGS, but aimed at preventing contagion, called a resolution fund.

The theoretical model of its activity is similar to the DGS one, the fundamental difference being in the values to be covered, since, in order to prevent contagion, the exposures under the guarantee are the interbank exposures, instead of deposits.

Here also, in order to prevent moral hazards and reduce the RF dimensioning, the coverage is typically partial.

With reference to simulations, this part of the safety net has an important impact, as it cuts out contagion for the small crises under the RF coverage. Instead, if the crisis is larger than the FR capabilities, contagion spreads out similar to the cases where no RF was set up, just reduced by the RF funds' value.

4.5 Computing Capital Coverage from Assets PD and Bank PD

Following Merton (1974), a loan is in default when its value falls below a value threshold, so when the loss overcomes the value C_i,

$$L_i > C_i \qquad\qquad (4.8)$$

Thus, the probability to default of the loan, APD, is the probability that losses are higher than C_i

$$APD = P(L_i > C_i) \qquad\qquad (4.9)$$

We can represent the value variation (losses) as the sum of two components, a common factor z and an idiosyncratic one, d:

$$L_i = \sqrt{R}z + \sqrt{1 - R}d \qquad\qquad (4.10)$$

so that

$$P(L_i > C_i) = P\left(\sqrt{R}z + \sqrt{1 - R}d > C_i\right) \qquad\qquad (4.11)$$

If L_i follows a standard normal distribution

$$C_i = N^{-1}(APD) \qquad\qquad (4.12)$$

where $N(x)$ is the normal standard cumulative distribution and $N^{-1}(x)$ is its inverse function.

When considering a bank, we do not refer to a single loan, but refer to a loan portfolio, where the diversifiable component is actually diversified. What actually threatens the bank stability is the undiversifiable component, this risk not being possible to neutralize as it affects all loans at the same time.

So, the attention must be focused on what capital value is needed for covering some specific probability of default for the bank.

This means that we have to look for the capital value that covers the probability of default of the loan when the common factor is at the stressed value.

If we want to have a preset value for the bank's probability of defaulting, BPD, the common factor must be set to $1 - BPD$; we can call the bank's probability of safeness, BPS_i.

So,

$$P(L_i > C_i | z = BPS_i) = P\left(\sqrt{R}\,N^{-1}(BPS_i) + \sqrt{1 - R}d > C_i\right)$$
$$(4.13)$$

Remembering that $C_i = N^{-1}(\text{APD})$ and $P(d > x) = N(x)$, we have

$$P(L_i > C_i | z = \text{BPS}_i) = N\left(\frac{N^{-1}(\text{APD}) - \sqrt{R}\, N^{-1}(\text{BPS}_i)}{\sqrt{1 - R}}\right)$$

(4.14)

Considering that loan provisions cover the expected losses $\text{APD} \times \text{LGD}$, the losses to be covered by capital are

$$C_i = \text{LGD} \times N\left(\frac{N^{-1}(\text{APD}) - \sqrt{R}\, N^{-1}(\text{BPS}_i)}{\sqrt{1 - R}}\right) - \text{APD} \times \text{LGD}$$

(4.15)

Now, referring to a bank loan's portfolio instead of a single loan (as in the Basel II framework), the capital requirement K_i of the bank i is given as the sum of the capital requirements dues for each loan of amount A_{in} to the firm n. This value is based on the probability of a default of the considered loan, APD_{in}:

$$K_i = \sum_n C_{in}(\text{APD}_{in}; \text{BPS}_i) \times A_{in}$$

(4.16)

In this way it is possible to compute the capital coverage needed for a preset bank PD given the assets PD.

4.6 Computing Banks Probability to Default from Capital Coverage and Assets PD

Within the Basel II framework, thus not considering contagion risks, the minimum capital requirement is set to cover 99.9% of the losses probability distribution. So, when a bank holds exactly the minimum required capital, the bank PD is 0.1%. But when, as typically happens, a bank holds more capital than required, it is not simple to quantify the bank's probability of defaulting.

Evidently, the higher the actual capital, the higher the part of the losses probability distribution coverage, so the lower the bank PD.

With the same approach described above, it is possible to compute the bank's probability of defaulting, BPD, as the part of the bank losses probability distribution not covered by its capital. It can be obtained just inverting the capital allocation formula, on the basis of the known APD_i and K_i, and numerically finding the value of BPD_i that verifies the equation:

$$\frac{K_i}{\sum_n A_{in}} = C_i\left(\overline{APD_i}, (1 - BPD_i)\right) \tag{4.17}$$

When, instead, contagion risks are considered, or the capital allocation formula cannot be applied, the approach must be different.

In fact, not only are contagion risks affected by more variables (interbank exposures) but mainly they depend on the system in which the bank is acting. If the other banks are weak and highly connected contagion risks, then default risks are higher independent of the idiosyncratic (without contagion) default risk of the considered bank.

So, the estimation of the "with contagion" default risk can better be based on Monte Carlo simulations.

In this case, the bank's probability of defaulting is obtained as the ratio between the number of defaults of the bank and the total number of simulations.

Recalling Table 2.6, we can just sum up the values in a column, thus having, for example, two defaults for bank 2 out of 10 simulations, and so a default rate of 20%. If, as often happens, the software only records the simulations where at least one bank defaults, it is important to remember that the default rate is obtained not as the ratio between the number of defaults and the number of lines in the matrix, but as the ratio between the number of defaults and the number of simulations performed for obtaining the matrix, including the "all zero" lines.

For further details, comparing the bank probability with default without contagion (which can also be obtained performing the same Monte Carlo simulation without contagion or setting to zero the interbank exposures) and with contagion, it is also possible to split the total default risk of the bank as the sum of the idiosyncratic risk plus the contagion risk:

$$BPD_i^T = BPD_i^c + BPD_i^{nc} \tag{4.18}$$

where BPD_i^T is the total probability of the bank defaulting (obtained from Monte Carlo simulations with contagion), BPD_i^{nc} is the bank's probability of defaulting without contagion (obtained either from the numerical inversion or performing a Monte Carlo simulation without contagion), and BPD_i^c is the total probability of the bank defaulting only due to contagion.

4.7 Risk Contributions and SiFis

One of the main issues related to systemic risk is in quantifying each single bank's contribution to systemic crises. This quantification is fundamental as a first step for understanding, and thus preventing, systemic crises.

In recent years, there has been growing interest in the topic, and the literature has analyzed it from both a theoretical and empirical perspective, and different approaches have also been developed.

One stream of research based the analysis on market data. These methods, relying on the behavior of each considered bank and of the whole market, try to extrapolate the effects of the system crisis on a single bank, or of single bank's distress on crises, so as to quantify the role of each bank in the system's riskiness.

In fact, two limits are evident for this approach. The first is due to the fact that crises are extreme events, and thus rare by definition, and as crises can have different causes and happen in different systems, the use of the information coming from this already limited number of cases is even more difficult.

The second problem is related to the actual market's capability of measuring the system risk. Backtesting for the capabilities of market-based measures to foresee banking systemic crises reveals that these measures have limited added value, at least for regulators. Zhang et al. (2015) proved that only DeltaCoVaR "consistently adds predictive power to conventional early warning models" and that the additional predictive power remains small, and it is not normally confirmed for other previous crises.

Thus, the use of simulations can be important, in approaching this problem, as these can yield all the values needed to analyze what can happen in systemic crises.

What is interesting is that the largest part of the measures developed for market data can also be applied to simulated data, so that the most sophisticated and effective models can be coupled with the data available from simulations, to develop an in-depth analysis of the mechanisms and determinants of the system riskiness and of each single bank's contribution to it.

4.7.1 Value at Risk (VaR)

The best-known and most commonly used measure within banking management is the value at risk (VaR). VaR was developed to obtain a comparable measure of risk for different banking activities. This measure proved to be a fundamental tool for risk management, even though it has some important limits.

The VaR at threshold s is defined as the loss value for the considered portfolio such that the probability of a higher loss is s% of cases.

VaR can be interpreted as the minimal amount of capital to be put back by the investor in order to preserve their solvency with a probability s%.

The strength of VaR relies on several important aspects:

1) VaR applies to any financial instrument.
2) It reports in money value an estimate of future events in converting the risk of a portfolio in a single value.

In real cases, the complexity of computational aspects and of the estimate of market probabilities, strong assumptions both on the dependence of financial instruments from risk drivers and on the probability distributions are needed to compute VaR.

When applying VaR to simulations, the problems related to the estimate of probability distributions are to be considered

Table 4.7 Reordered excess losses.

	Bank 1	Bank 2	Bank 3	Bank 4	Bank 5	Total
Iteration 3						0
Iteration 5						0
Iteration 6						0
Iteration 8						0
Iteration 9				2.30		2.30
Iteration 1	28.89					28.89
Iteration 4			121.67		0.24	121.90
Iteration 2			191.07	4.49		195.56
Iteration 7	59.89	41.52	242.43	17.11		360.95
Iteration 10	249.03	65.59	373.88	10.40		698.89
Total	337.80	107.11	929.03	34.30	0.24	

when modeling the random shocks and correlations, so that the simulation results can be a correct representation of the expected distribution of results.

In this case, the VaR computation is simple. Starting from a banking system simulation, as in Table 2.7, we just have to reorder the values and select the considered threshold value. In this example, the $VaR_{90\%}$ corresponds to the 9th value, thus 360.95 for the system total and 59.89 for bank 1.

Note that, generally speaking, the total reordering does not give the proper reordering of each column, thus for each of the bank values. In this example, bank 4's values are evidently not correctly ordered, so if the aim is to compute the VaR of bank 4, the reordering must be done on the basis of the column values.

But, what is more important is the fact that for the same reason, the sum of each bank VaR does not match the total system VaR at the same threshold. In this example, the $VaR_{90\%}$ values for each bank are 59.89, 41.52, 242.43, 10.40 and 0, summing up to 354.24, while the system $VaR_{90\%}$ is 360.95.

It is evident from the example that the system VaR cannot be computed as the sum of the VaR for the single components (banks) due to the only partial correlation between them, and so VaR is not an additive measure.

In fact, VaR is not a coherent measure,[1] since, although it satisfies several important properties such as translation invariance, positive homogeneity, and monotonicity, it fails to satisfy the subadditivity property.

Another of the VaR limitations is that it does not consider at all what happens after the threshold, so VaR will not signal if the tail losses are really high.

A simple example can highlight the consequences of this limit.

Consider a portfolio of US$1000, with a maximum loss level of US$100 and suppose that the worst 5% cases on 1 month are all of this maximum loss. $VaR_{95\%}$ on this time horizon would then be US$100.

Consider now another portfolio of the same value that invests also in positions that allow for a potential unbounded maximum loss, such that its VaR is still US$100 on the time horizon, but the 5% worst-case losses range from US$100 to a highly higher value.

According to the $VaR_{95\%}$, the two portfolio risks are equivalent, which is evidently not the case.

4.7.2 Expected Shortfall (ES)

An alternative measure of risk aimed at measuring the tail risk, is the expected shortfall (ES) (see Acerbi *et al.* (2001)). Expected shortfall (or conditional VaR or tail VaR) has been characterized as the smallest coherent and law invariant risk measure to dominate VaR.

ES accounts for the expected value of losses after the threshold, thus focusing on the tail risk.

As the ES represents the expected loss for a given portfolio in the worst $(1 - q)$ share of cases, it is a more appropriate measure

---

1 For a definition of coherent risk measure, see Artzener *et al.* (1999) Definition of coherent measures of risk. *Mathematical Finance*, **9**, July, pp. 203–228.

of systemic risk than the VaR, which only represents the minimum loss in the worst $(1 - q)$ share of cases, thus in the "normal" states of the system.

With reference to the example in Table 4.7, and referring, for evidencing the measure, to the $VaR_{50\%}$, we record a value of 2.30, the worst value of the best five out of 10 cases. Nothing is considered above this threshold. Instead, the $ES_{50\%}$ would register a value of 281.238, obtained as the simple average of the values above the threshold.

4.7.3 Conditional Value at Risk (CoVaR)

Another VaR limit is that it focuses on the risk of an individual institution in isolation.

Conditioning VaR and ES, several further measures were proposed.

Adrian and Brunnermeier (2011) proposed a different approach called CoVar, and its specification DeltaCoVaR.

Institution i CoVaR relative to the system is defined as the VaR of the whole financial sector conditional on institution i being in distress (as measured by capital losses). The difference between the CoVaR and the unconditional financial system VaR, Delta-CoVaR, captures the marginal contribution of a particular institution (in a noncausal sense) to the overall systemic risk.

CoVar estimates the risk contribution of a single institution as the value at risk of the financial system conditional on the distress of an individual institution, capturing the correlation effects between the distress of a single bank and the whole system.

CoVar is implicitly defined as follows:

$$\Pr\left(L > \text{CoVar}_q^i|_{L_i > t_i}\right) = q \tag{4.19}$$

where q is the preset probability level and t_i is the distress threshold for bank i.

A further step to measure the contribution of a single institution to the overall systemic risk is proposed in the same paper called ΔCoVar.

ΔCoVar is defined as the difference between the CoVar conditioned to the distress of an individual bank and the CoVar conditioned to the median state of the same bank,

$$\Delta CoVar_q^i = CoVar_q^i - CoVar_q^{median} \tag{4.20}$$

More precisely, DeltaCoVaR of bank i is the difference between the VaR of the financial system when bank i is in distress and the VaR of the financial system during a "normal" state of the considered bank.

Delta CoVaR focuses on the contribution of each institution to overall system risk, and thus can be an important tool for studying the risk spillover effects across the whole financial network.

The authors have deliberately not specified how to estimate the CoVaR measure.

With reference to simulations, the measure can be computed selecting the cases where the considered bank reports losses from the cases where no losses are computed for the bank.

Starting from the same data in Table 2.7, and referring to bank 4, we can consider the cases where bank 4 reports losses from the cases where the bank is not suffering losses.

We find that the $CoVaR_{80}\%$ of the system conditional on bank 4 being in distress is 195.56, the third value for the system out of 4, while the $VaR_{80}\%$ in the remaining six cases where bank 4 suffers no losses is 28.89.[2] Thus, the DeltaCoVar is of 166.67.

CoVar can also be declined in other ways. For example, DeltaCoVaR$_{ji}$ captures the increase in risk of individual institution j when institution i falls into distress. To the extent that it is causal, it captures the risk spillover effects that institution i has on institution j. Of course, it can be that institution i distress causes a large risk increase in institution j, while institution j causes almost no risk spillovers onto institution i. That is, there is no reason why DeltaCoVaR$_{ji}$ should equal DeltaCoVaR$_{ij}$.

---

2 More precisely, it is somewhere between 28.89 and 121.90.

Table 4.8 Simulations reordered on total losses selected by default/ nondefault of bank 4.

	Bank 1	Bank 2	Bank 3	Bank 4	Bank 5	Total
Iteration 3						0
Iteration 5						0
Iteration 6						0
Iteration 8						0
Iteration 1	28.89					28.89
Iteration 4			121.67		0.24	121.90
Iteration 9				2.30		2.30
Iteration 2			191.07	4.49		195.56
Iteration 7	59.89	41.52	242.43	17.11		360.95
Iteration 10	249.03	65.59	373.88	10.40		698.89
Total	337.80	107.11	929.03	34.30	0.24	

In the example in Table 4.8, we can compute the $CoVar_{3,4\ 80\%}$, i.e., the third worst case for bank 3 out of the four cases where bank 4 records some losses, and compare it with the same measure computed when bank 4 suffers no losses, for obtaining the $DeltaCoVar_{3,4\ 80\%}$. In this case it is 242.43, while the $DeltaCoVar_{4,3\ 80\%}$ is 10.40.

Evidently, these results depend on the threshold selected for distinguishing "normal" cases from "distress" cases. It is also possible to base the same measures on cases where the bank is not in actual default, but just near it, or suffers a significant undercapitalization, or other. In any case, the results are determined not only by the correlation used as input, but also by capitalization, assets riskiness, interbank exposures, and dimension.

Another attractive feature of CoVaR is that it can be easily adopted for other co-risk measures, such as the co-expected shortfall (Co-ES). Expected shortfall has a number of advantages relative to VaR and can be calculated as a sum of VaRs.

The $CoES_q^i$ is the expected shortfall of the financial system conditional on $X^i \leq VaR_q^i$ of institution i. That is, $CoES_q^i$ is defined by the expectation over the q-tail of the conditional probability distribution:

$$EX^{\text{system}} | X^{\text{system}} \leq CoVaR_q^i \qquad (4.21)$$

Institution i's contribution to $CoES_q^i$ is simply denoted by

$$
\begin{aligned}
DeltaCoES_q^i &= EX^{\text{system}} | X^{\text{system}} \leq CoVaR_q^i \\
&\quad - EX^{\text{system}} | X^{\text{system}} \leq VaR_q^{\text{system}}
\end{aligned} \qquad (4.22)
$$

4.7.4 Marginal Expected Shortfall (MES)

Acharya *et al.* (2010) in a similar approach proposed the marginal expected shortfall as a measure of systemic risk.

The marginal expected shortfall or systemic expected shortfall (SES) measures the systemic risk of an individual bank as its propensity to be undercapitalized when the system as a whole is undercapitalized. They defined their measure by the worst 5% of equity returns measured from a historical time series with a daily frequency.

Acharya *et al.* (2010) showed a way to identify how each financial institution's contribution to overall systemic risk can be measured. They proposed the systemic expected shortfall, that is, the expected shortfall of a bank conditional to the distress of the system as a whole:

$$SES_i = E(L_i|_{L>T}) \qquad (4.23)$$

where T is the systemic crisis threshold in terms of VaR, L_i is the excess loss of bank i (over the capital buffer, a default event being defined as $L_i > 0$), L is the excess loss in the whole banking system $L = \sum_i L_i$, and SES thus refers to that part of systemic risk that affects each single bank.

With reference to simulations, the risk contribution of each bank in these crises can be computed as the column average of the selected crises.

Table 4.9 Selected worst 50% crises.

	Bank 1	Bank 2	Bank 3	Bank 4	Bank 5	Total
Iteration 1	28.89	0	0	0	0	28.89
Iteration 4	0	0	121.67	0	0.24	121.90
Iteration 2	0	0	191.07	4.49	0	195.56
Iteration 7	59.89	41.52	242.43	17.11	0	360.95
Iteration 10	249.03	65.59	373.88	10.40	0	698.89
Average ($ES_{50\%}$)	67.56	21.42	185.81	6.40	0.05	281.24

Starting from the values in Table 2.7 and considering two different thresholds (Tables 4.9 and 4.10), we can see that the risk contributions are different when considering different thresholds.

It is worth noting that while normally the risk contributions are higher for the more severe crises, in some cases (bank 5) the risk contribution is lower. This happens when some (small) bank is not strongly linked to the rest of the system, both in correlation and in interbank exposure terms, so that it only defaults in small crises, and is not affected by contagion.

Another important note is that single risk contributions are derived first by selecting the most severe crises for the system, and then computing the average for the single bank. Computing the ES for the single bank just selecting the worst cases for that bank will not give an additive measure of the system risk (e.g., bank 5 $ES_{70\%}$ is not zero, as the loss in iteration 4 is included).

Table 4.10 Selected worst 30% crises.

	Bank 1	Bank 2	Bank 3	Bank 4	Bank 5	Total
Iteration 2	0	0	191.07	4.49	0	195.56
Iteration 7	59.89	41.52	242.43	17.11	0	360.95
Iteration 10	249.03	65.59	373.88	10.40		698.89
Average ($ES_{70\%}$)	102.97	35.70	269.12	10.67	0	418.46

4.7.5 Shapley Values

Recently, some papers (Tarashev *et al.* 2009; Drehmann and Tarashev, 2013) used the Shapley value for assessing individual risk contribution based on simulations. The Shapley value (see Shapley, 1953) is one of the most important solution concepts in cooperative games, that is, in contexts where the competition is between coalitions of players, rather than between single players (e.g., voting games). The Shapley value has a number of desirable properties, such as additivity, symmetry, and zero player.

Additivity ensures that it distributes exactly the total benefit to all players, without resulting in any surplus or deficit. Symmetry ensures that two players with the same characteristics receive the same share of the overall value. Thus, the gains from cooperation between any two players are divided equally between them. Zero player means that no payoff is assigned to a player who makes no contribution to any subgroup.

The Shapley value for a player in a game can be defined as the average of that player's marginal contributions to every possible coalition. Its computation relies on the evaluation of the outcome for each possible coalition of players, and so the risk carried by each subsystem of banks in our case. Table 4.11 reports an example of computing the Shapley value, where each of the "subgroup outcomes" is the risk measure obtained by means of a Monte Carlo simulation.

The results in Table 4.11 show several important characteristics of Shapley values.

First, as the measure is additive, summing the Shapley values of each bank we have the risk value computed for the whole system $(26.08 - 2.92 + 481.42 + 87.42 = 592)$.

Second, its value depends on two components: the outcome of the single player and the nonlinearities in the outcomes of the subsystems. If the risk were just additive, so that the risk brought by the subsystem A,D were of 104 $(20 + 84)$ and so on with all the subgroups, the single player outcomes would be equal to the Shapley values. Thus, it is worth computing the risk contributions in this way only when the simulation model includes contagion. This also means that if the outcome is measured

Table 4.11 Example of Shapley value computation for a sample of four banks.

Subgroup	Subgroup outcome	Marginal contribution to subgroup of bank			
		A	B	C	D
A	20	20			
B	1		1		
C	477			477	
D	84				84
A,B	21	20	1		
A,C	505	28		485	
A,D	106	22			86
B,C	478		1	477	
B,D	85		1		84
C,D	561			477	84
A,B,C	500	22	−5	479	
A,B,D	107	22	1		86
A,C,D	604	43		498	99
B,C,D	560		−1	475	82
A,B,C,D	592	32	−12	485	92
Contribution to subgroups of one player		20.00	1.00	477.00	84.00
Average contribution to subgroups of two players		23.33	1.00	479.67	84.67
Average contribution to subgroups of three players		29.00	−1.67	484.00	89.00
Contribution to subgroups of four players		32.00	−12.00	485.00	92.00
Shapley values		26.08	−2.92	481.42	87.42

as the $VaR_{90\%}$ of the considered subsystem, and contagion only starts in the last 5% of the losses distribution, computing the Shapley values is just a waste of time, being equal to the single bank riskiness. Conversely, the factors affecting nonlinearities, such as the share of interbank exposures, the exposure to common risk factors, and the low capitalization, in determining a higher contagion risk, also increase the systemic importance.

Another important feature of the Shapley value is that it is also possible to have negative risk contributions.

In fact, if a bank has a strong capitalization and high exposure in lending to banks but no borrowing from other banks, it can happen (bank B in the example in Table 4.11) that its high loss-absorbing capacity lowers the contagion risk, as the bank absorbs some losses, but its resilience means that it does not default. When, instead, that bank is not in the system, its share of interbank losses is shared between the remaining banks, and this can make another bank default by contagion. Consequently, the marginal contribution of the bank to systemic risk proves to be negative.

On the other hand, the actual use of this approach is strongly limited by its computational complexity: In a set of N players, the number of all possible subgroups is equal to 2^N, so, for example, a game with four players requires the computation of Monte Carlo simulations for 16 different subgroups, while a 20-player system requires the computation of one set of Monte Carlo simulations for each of 1,048,575 subgroups.

4.7.6 The Leave-One-Out Approach

In order to overcome the computational limits of Shapley values, one can use another method, used but not specifically analyzed in Gauthier *et al.* (2012) for computing incremental VaR, ΔCoVar, and marginal expected shortfall, with the aim of verifying if a capital allocation based on these measures will lead to a safer banking system, and more specifically developed in Zedda and Cannas (2016), called leave-one-out.

This approach is based on the idea that the contribution of each bank to the system can be obtained by comparing the banking system performances with the performances of the same banking system when excluding the considered bank. This case seems actually more realistic, with reference to banking systems. In fact, banks usually do not act as part of subgroups, which is instead the case, for example, in voting games, where each actor joins parties, and calculating the value added in each of all possible parties represents the standard framework the Shapley value is designed for.

It is, instead, actually possible for the authorities to ring-fence a risky bank from the interbank market, for example, for preserving the system's safeness.

From another point of view, the LOO proves to be a first-order approximation of the Shapley value. With reference to the example in Table 4.11, LOO keeps the first component, contribution to subgroups of 1 player, and the last one, the marginal contribution to the whole system. While the first component carries the single bank riskiness, the contribution to the whole system keeps the nonlinearities in the system.

In exchange for this approximation, LOO allows risk contributions for actual banking systems to be computed, as its computation time is proportional to the number of banks in the system, and this approach is not limited by the system dimension.

Being based on balance sheet data and simulations, this method allows the risk contribution of all banks, not only the listed ones, to be quantified, which is fundamental for supervision and regulation purposes. In addition, it can be applied also to banks that were not in distress, so no market data are available on their behavior during crises.

With reference to the method characteristics, one important feature of this method is that it makes a distinction between correlation and contagion, so there are cases when banks are sensitive to the crisis (or its determinants) and cases when the effect is due to the exposures.

This distinction can be explained by means of a simple example. Consider the case of a small bank with high correlation with the crisis determinants. In case of an exogenous common shock, both the bank and the system will experience a high impact on its market value.

In this case, the measures based on a comprehensive correlation, which includes contagion, such as CoVaR (and ΔCoVaR) and SES, will capture this comovement even if the impact of the considered bank on the system is small.

Instead, the LOO measure, comparing the performances of the system when a bank is included or not, will give a measure of how the bank is contributing to the system riskiness, assessing for the actual role: The bank is affected by the crisis but not inducing the crisis start.

Instead, a large and highly exposed bank will trigger contagion. In this case, including or not including the considered bank will make an important difference, because when the bank is excluded, contagion will start no more. LOO accounts for this difference, while other measures will just signal correlation in losses due to contagion. So, this measure allows one to distinguish between cases where the bank suffers from the crisis without causing it (correlation) and cases where the bank's presence has important effects on the crisis dimension (contagion).

Another fundamental feature of this approach is that it allows the risk contribution of each bank to be split into two components, namely, the stand-alone risk (expected primary losses) and the contagion component, computed as the expected variation in systemic losses, due to the bank's linkages to the other banks in the system. From this, one can evaluate the effects of the interbank linkages on the system safeness, autonomously from the single banks' safeness, with important insights into macroprudential regulation.

The method can be applied by means of different measures of risk, but expected shortfall (ES) seems to be more suitable for evaluating the tail risk related to systemic crises.

The difference between the expected shortfall for the entire banking system (L) and the ES of the same system when leaving the considered bank h out of the system ($L^{(h)}$) can be split as the sum of two components, namely, the stand-alone risk contribution of the bank, that is, the loss of the h bank (L_h) and the "contagion risk contribution" of the h bank to the system (Sys_h):

$$L - L^{(h)} = L_h + \text{Sys}_h \tag{4.24}$$

Sys_h is the value of the differential losses for the system when the bank h and the rest of the system are connected, thus accounting for the loss transmission role of the bank.

This contagion risk contribution is normally positive; there are indeed cases where Sys_h is negative, signaling a loss barrier effect that the bank has in the system.

As the sum of the contagion risk contributions, $\sum_h Sys_h$ does not match the system total, to yield an additive measure; this component must be rescaled as in Huang *et al.* (2011) as follows:

$$Sys_h^* = Sys_h \times \frac{L - \sum_h L_h}{\sum_h Sys_h} \tag{4.25}$$

Finally, the LOO risk contributions are the sum of the stand-alone component L_h and the rescaled contagion risk component Sys_h^*.

Some interesting insights are reported in the following tables.

As Table 4.12 shows, different thresholds make different contributions for the same banks, so bank 7 and bank 8, going from the first column to the last one, have opposite trends: While bank 7 raises from 15.9 to 28.1%, bank 8 lowers its role from 27.7 to 18.0%.

Table 4.12 LOO contribution shares by crisis probability level.

Leave-one-out	All	99.900%	99.950%	99.990%	99.999%
French Bank 1	0.5%	0.4%	0.2%	0.2%	0.2%
French Bank 2	−1.0%	−1.0%	−0.8%	−0.8%	−0.8%
French Bank 3	11.7%	11.2%	7.5%	6.7%	8.2%
French Bank 4	0.8%	0.6%	0.3%	0.3%	0.3%
French Bank 5	2.6%	2.4%	1.4%	0.8%	0.6%
French Bank 6	27.1%	27.6%	30.9%	34.1%	34.0%
French Bank 7	15.9%	16.4%	18.8%	23.7%	28.1%
French Bank 8	27.7%	27.6%	26.3%	22.4%	18.0%
French Bank 9	14.8%	14.9%	15.3%	12.6%	11.4%
Total	100.0%	100.0%	100.0%	100.0%	100.0%

Source: Zedda *et al.*, 2016. Reproduced with permission of SSRN.

Table 4.13 Leave-one-out risk contributions and Shapley values, 99.99% threshold.

Banks	LOO contributions (th €) $Sys_h^* + L_h$	Shapley values (th €)	LOO unitary contributions	Unitary Shapley values
French Bank 1	180,130	104,838	7.17	4.17
French Bank 2	(780,183)	(805,117)	(22.46)	(23.18)
French Bank 3	6,790,660	7,191,800	33.72	35.71
French Bank 4	290,202	100,132	9.84	3.39
French Bank 5	811,717	869,701	9.72	10.41
French Bank 6	34,339,534	34,894,867	18.97	19.27
French Bank 7	23,919,104	24,177,345	14.17	14.32
French Bank 8	22,521,720	21,992,857	20.02	19.55
French Bank 9	12,658,880	12,205,341	10.43	10.05
Total	100,731,765	100,731,765		
Correlation	0.9996370		0.988930	

Source: Zedda *et al.*, 2016. Reproduced with permission of SSRN.

With reference to the Shapley values, from Table 4.13 we can see that total LOO contributions and Shapley values are very similar, so the correlation between the two is about 99.96%. Even neutralizing the dimension effect, comparing unitary LOO contributions (risk contributions on total assets) with unitary Shapley values, the values are remarkably similar, and the correlation remains very high, near 98.9%. This means that the first-order decomposition of the system, leaving one out, is the most important, accounting for the main part of the systemic contribution captured by the Shapley values. Further tests on different samples confirmed the high correlation values.

While the stand-alone component seems mainly determined by RWA (positive) and capital (negative), the contagion component is mainly linked to interbank exposures, keeping the capital barrier role always significant (Table 4.14). In all

Table 4.14 Unitary stand-alone component of risk contribution, regression coefficients, and significance.

	L_h/TA 99.900%		L_h/TA 99.950%		L_h/TA 99.990%		L_h/TA 99.999%	
const	0.0208	***	0.0335	***	0.0182		0.0293	
Ln (TA)	−0.0025	***	−0.0039	***	−0.0021		−0.0032	
Ln (TA)sqr	0.0001	***	0.0001	***	0.0001		0.0001	
TRC/TA	−0.0195	***	−0.0277	***	−0.0638	***	−0.1256	***
RWA/TA	0.0040	***	0.0056	***	0.0125	***	0.0254	***
IB_A/TA	−0.0006		−0.0010		−0.0017		−0.0038	
IB_D/TA	0.0013	**	0.0017	**	0.0030	*	0.0020	
R^2	0.602		0.598		0.575		0.537	

Source: Zedda *et al.,* 2016. Reproduced with permission of SSRN.

estimations, the significant variables have rising coefficients as the crisis size increases.

In all the estimations, even when considering the unitary risk contributions (Tables 4.15 and 4.16), a higher capitalization

Table 4.15 Unitary contagion component of risk contribution, regression coefficients, and significance.

	Sys_h^*/TA 99.900%		Sys_h^*/TA 99.950%		Sys_h^*/TA 99.990%		Sys_h^*/TA 99.999%	
const	−0.0302		−0.0553		−0.3797		−1.4213	*
Ln (TA)	0.0032		0.0058		0.0404		0.1444	*
Ln (TA)sqr	−0.0001		−0.0002		−0.0011		−0.0036	*
TRC/TA	−0.1074	***	−0.1854	***	−0.4309	***	−0.8289	***
RWA/TA	0.0028		0.0043		0.0128		0.0782	*
IB_A/TA	0.0625	***	0.1113	***	0.3078	***	0.7512	***
IB_D/TA	0.0159	**	0.0285	**	0.0756	**	−0.1720	***
R^2	0.515		0.525		0.588		0.705	

Source: Zedda *et al.,* 2016. Reproduced with permission of SSRN.

Table 4.16 Unitary total LOO risk contribution, regression coefficients, and significance.

	LOO/TA 99.900%		LOO/TA 99.950%		LOO/TA 99.990%		LOO/TA 99.999%	
const	−0.0094		−0.0219		−0.3614		−1.3920	*
Ln (TA)	0.0007		0.0019		0.0383		0.1412	*
Ln (TA)sqr	0.0000		0.0000		−0.0010		−0.0035	
TRC/TA	−0.1270	***	−0.2131	***	−0.4947	***	−0.9545	***
RWA/TA	0.0067		0.0099		0.0253		0.1036	**
IB_A/TA	0.0619	***	0.1103	***	0.3061	***	0.7474	***
IB_D/TA	0.0171	**	0.0302	**	0.0786	**	−0.1699	***
R^2	0.513		0.524		0.588		0.707	

Source: Zedda *et al.*, 2016. Reproduced with permission of SSRN.

results in an important reduction in the systemic risk contribution, the coefficients being −0.0195 for the stand-alone component and −0.1074 for the contagion one at the 99.9% threshold, and −0.1256 and −0.8289, respectively, for the 99.999% threshold.

These results confirm the importance of exposures, but mainly of capitalization for determining the risk contribution.

4.7.7 Starting and Fueling Contagion: Risk Contribution Roles

As described in Section 1.4, contagion is the realization of two risks: the risk that a bank can default, and the risk that the initial default spillover spreads through the interbank channel, inducing other banks to default.

Accordingly, the simulation mechanism presented in Section 4.2 is always based on a first step based on idiosyncratic risk for determining whether some bank defaults are due to specific losses and, in case of some bank defaults, a contagion mechanism.

As the distinction between the two risks is reproduced by the simulation mechanism, and all the intermediate values are available, it is possible to verify if the two risks are linked to the same determinants, or to the same banks. In fact, if the two roles are distinct, so that some banks are typically "lighter" of the crisis while others act as "fuel," the supervision and regulation interventions to prevent a systemic crisis and its consequences can be differentiated, specifically tailored for each bank, and evidently more effective.

In applied terms, the role distinction relies on a selection of the simulated crises, selecting the cases where contagion started or not, and comparing the "contagion" simulations with the "no contagion" ones. In this way, it is possible to see which idiosyncratic crisis has determined contagion to start, and analyze which of the subjacent variables, either bank specific (assets riskiness, capitalization, interbank linkages, etc.) or exogenous shocks (common or idiosyncratic factor), are determinant for contagion to start.

One interesting experiment in this line is in Zedda *et al.* (2012c). The exercise is developed based on correlated risk factors, comparing simulations with and without contagion (maintaining the same random seed in each run), to distinguish between "primary" and "induced" defaults and fragility. Different systemic risk contribution measures are computed to capture the total shortfall expected for the system, given the failure of a single institution. The analysis is tested on the Danish banking system for 2010.

Following the CoVar approach, conditional risk contributions can be obtained as the VaR of the whole system conditional to the default of bank i:

$$\Pr\left(L > \text{CoVar}_q^i\big|_{Li>0}\right) = q \qquad (4.26)$$

In this case, the condition is on the cases where bank i defaults or not, Li measuring the value of losses in case of default, so $L > 0$ signals a default, and $L = 0$ means no default.

ΔCoVar becomes

$$\Delta\text{CoVar}_q^i = \text{CoVar}_q^i\big|_{Li>0} - \text{CoVar}_q^i\big|_{Li=0} \qquad (4.27)$$

Similarly, ΔMES is computed starting from the expected short-fall for the system conditional to the distress of bank i as

$$\text{MES}^i_T = E\left(L|_{Li>0,L>T}\right) \tag{4.28}$$

and the differential value ΔMES is given by

$$\Delta\text{MES}^i_T = E\left(L|_{Li>0,L>T}\right) - E\left(L|_{Li=0,L>T}\right) \tag{4.29}$$

where T is the systemic loss threshold and, as for the ΔCoVar, the "normal case" is the no-default status of institution i.

In both cases, ignition risk contributions can be calculated as the expected value of a large crisis, conditional on the primary default of the considered bank.

Results suggest that the methodology could be effective in distinguishing the different roles played by banks in crises. In Figure 4.4, some cases register overall contributions and primary contributions with similar values, identifying potential "lighters," while other cases signal large differences between

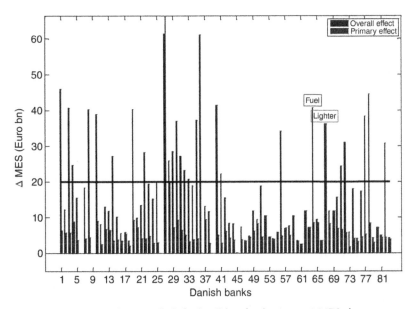

Figure 4.4 ΔMES due to all defaults (blue bar) versus ΔMES due to primary defaults (red bar). *Source:* Zedda, 2012. Reproduced with permission of European Union.

overall and primary effects, thus signaling a potential "fuel" role in the crisis.

4.8 The Regulator's Dilemma

One interesting problem that can be analyzed by means of simulations is the so-called Regulator's Dilemma.

As already considered when describing the role of correlations, the main risk one banking system may face is a systemic crisis, where contagion starts and affects a large part of the system.

As verified by many studies, such as Steinbacher *et al.* (2016), the main source of a systemic crisis relies on common shocks. Thus, the level of correlation between banks, or between banks and macro factors, is crucial.

On the other hand, we can consider that the same relevance is in the correlation between a bank's assets with reference to the single bank stability. The more a bank's assets are diversified, the more stable its results, and thus the lower the risk of a bank failure.

When considering the two layers of correlation from the regulator point of view, one question arises: Is it better to have specialized banks, with a higher concentration of risk factors within each bank, so as to have a higher diversification between banks, or the safer solution for the system in having a deep diversification within each bank portfolio but lower diversification between banks?

In the first case, single banks are weaker, but as each risk factor is significant only for a fraction of the system, the system as a whole has lower probabilities to be hit seriously by contagion. In the second case, banks are stronger as single, but all the banks are exposed to the same risk sources.

Which structure is safer for the system?

Is the optimal solution for the single bank different from the optimal solution for the system?

Beale *et al.* (2011) analyzed this problem in a theoretical modeling, verifying that while the single bank optimum relies on the maximum diversification within the bank, the lower probability of joint failure, the system optimum is in the full specialization when each bank invests in one different asset.

These results give an important clue to the problem, although in the simplified model some important variables, such as the correlation between assets, interbank exposures, and capital levels, are not considered. In addition, the evaluation of what can happen in a more complex and "realistic" banking system, where specialization can include economic sectors, regions, and time horizons, the risk sources are in someway correlated, and all the possible complexity is considered, suggesting that more analyses might be warranted.

Appendix: Software References and Tools
By Simone Sbaraglia

Here we will include references on how to implement the models and some examples.

The precision of Monte Carlo simulation results mainly depends on the number of simulations performed.

Thus, the implementation of high-level software is not the best solution, as it results in a much slower execution, and therefore in wider intervals of confidence in its results. Instead, programming in a low-level language allows for faster runs and thus the ability to perform a higher number of iterations, which yields higher-quality results.

In the following, we present a C language implementation of the models, together with the description needed to execute and possibly modify and improve the software.

Let us describe here a sample implementation of the algorithms outlined in the book. The high-level structure of the source code is an implementation of a Monte Carlo algorithm, whose simulation accuracy highly depends on the number of iterations performed. Therefore, it would be inefficient to write the code in a high-level interpreted programming language. The samples are therefore provided in the C programming language.

The structure of the source code is as follows:

main function {

1) interpret command line args
2) read input files

Banking Systems Simulation: Theory, Practice, and Application of Modeling Shocks, Losses, and Contagion, First Edition. Stefano Zedda.
© 2017 John Wiley & Sons, Ltd. Published 2017 by John Wiley & Sons, Ltd.

3) initialize data structures
4) allocate data structures
5) compute correlation matrix
6) Monte Carlo external loop, proceeds until the number of defaults equals nTotDefaults. At each iteration of the loop:
   6.1 generate random bank losses
   6.2 simulate interbank contagion
7) at the end of Monte Carlo loop dump all output files

}

1) **Interpret command line args**

The command line options provided are the following:

-nointerbank	do not simulate interbank contagion
-dumpbankloss	output the matrix of nbanks columns by ntotdefaults rows containing the excess losses for each bank in each simulation with at least one default (full mapping)
-dumpecontrib	output a line vector with the risk contribution for each bank above of the "large loss limit" set
-dumpecloss	output a line vector with the excess losses for each simulation
-randomseq	the random numbers generation is different in every set of simulations

Furthermore, the software expects one input file with the following contents: assets PD, deposits, capital, total assets, interbank debts, interbank credits, and correlation.

This version of the software includes the distinction of the losses above a preset threshold.

When launching the execution, the command line must contain the input file name and the "large loss limit" value. Launching the simulation process without reference to any threshold (so all losses) is as simple as setting the "large loss limit" value to zero.

```
#include <stdio.h>
#include <stdlib.h>
```

```
#include <unistd.h>
#include <math.h>
#include <string.h>

#include <gsl_randist.h>
#include <gsl_cdf.h>
#include "common.h"

#include <gsl_errno.h>
#include <gsl_matrix.h>
#include <gsl_rng.h>

#define NARGS 2
#define MAXSTRLEN 256
#define NCOLS 7

void Usage(char *s) {
   fprintf(stderr, "Usage: %s [-nointerbank | -
dumpbankloss | -dumpecontrib |-dumpecloss | -
randomseq] inputfile LargeLossLimit\n", s);
   exit(-1);
}

double R(double x) {
   return(0.12*(1-exp(-50*x))/(1-exp(-50))+0.24*
(1-(1-exp(-50*x))/(1-exp(-50)))));
}

double ba(double x) {
   return(pow(0.11852-0.05478*log(x), 2));
}

double regcap(double x, double y, double k) {
   double a;

   a = (k*gsl_cdf_ugaussian_P(
             pow(1-R(x), -0.5)*gsl_cdf_ugaussian_Pinv(x)+
             pow(R(x)/(1-R(x)), 0.5)*y) - x*k)*
         pow(1-1.5*ba(x), -1)*1.06;
   return(a);
}
```

```c
void printDoubleArray(char *s, int n, double *p) {
    int i;

    if (s!=NULL) {
        printf("-->ARRAY %s\n", s);
    }
    for (i=0; i<n; i++) {
        printf("%f\n", p[i]);
    }
}

int main(int argc, char *argv[]) {
    double rho=0;
    double LGD=0.45;
    double BankLossCap = 0;
    char *filename = NULL;
    char *outfnameprefix1 = "Bankloss";
    char *outfnameprefix2 = "DIScharged";
    char *outfnameprefix3 = "Econtrib";
    char *outfnameprefix4 = "EcLoss";
    char *outfname1;
    char *outfname2;
    char *cont = "n";
    FILE *fp = NULL;
    FILE *ofp1 = NULL;
    FILE *ofp2 = NULL;
    FILE *ofp3 = NULL;
    FILE *ofp4 = NULL;
    char buffer[MAXSTRLEN];
    double **data = NULL;
    double **BankLoss = NULL;
    double *RContrib = NULL;
    double *EcLoss = NULL;
    double *DIScharged = NULL;
    double *ECcharged = NULL;
    double *market = NULL;
    double *corr = NULL;
    int *BankDefaulted = NULL;
    int *BankDefaultedOld = NULL;
    double InterbankUnitaryLoss = 0;
    double Baseloss = 0;
    int i, b, j, nDefaults;
```

```c
int nbanks = -1;
int niter = -1;
int hadDefault;
int hadDefaultInterbank = 0;
int numDefaultedBanks = 0;
int allDefaulted;
double InterBankLoss, InterBankDefault;
int argnum=1;
int intsize1 = 0;
int intsize2 = 0;
int intsize3 = 0;
int intsize4 = 0;
int interbanksimulation = 1;
int dumpbankloss = 0;
int dumpecontrib = 0;
int dumpecloss = 0;
int nTotDefaults = 100000;
int LargeLossLimit = 0;
double tmp=0;
const gsl_rng_type * Ra;
gsl_rng * r;
gsl_rng_env_setup();
Ra = gsl_rng_default;
r = gsl_rng_alloc (Ra);

/* checking number of args */
if (argc < NARGS + 1) {
   Usage(argv[0]);
   exit(-1);
}

while (argnum < argc-NARGS) {
  if (!strcmp(argv[argnum], "-nointerbank")) {
     argnum++;
     interbanksimulation = 0;
     fprintf(stderr, "Interbank Simulation Disabled\n");
  } else if (!strcmp(argv[argnum], "-dumpbankloss")) {
     argnum++;
     dumpbankloss = 1;
     fprintf(stderr, "Dumping BankLoss Enabled\n");
  } else if (!strcmp(argv[argnum], "-dumpecloss")) {
```

```
          argnum++;
          dumpecloss = 1;
          fprintf(stderr, "Dumping Economic Losses
  Enabled\n");
      } else if (!strcmp(argv[argnum], "-dumpecontrib")) {
          argnum++;
          dumpecontrib = 1;
          fprintf(stderr, "Dumping Economic Risk
  Contributions Enabled\n");
      } else if (!strcmp(argv[argnum], "-randomseq")) {
          argnum++;
          srandom(time(NULL));
          fprintf(stderr, "Init random seed\n");
      } else {
          Usage(argv[0]);
      }
    }

    /* reading command line */
    filename = argv[argnum];
    if ((fp = fopen(filename, "r")) == NULL) {
        fprintf(stderr, "could not open file %s for reading
  \n", filename);
        exit(-1);
    }

    LargeLossLimit = atof(argv[argnum+1]);
    if ((LargeLossLimit < 0)) {
        fprintf(stderr, "LargeLossLimit cannot be
  negative: %s\n", LargeLossLimit);
        exit(-1);
    }

    intsize1 = 10;
    intsize2 = floor(log10(nTotDefaults)) + 1;
    intsize3 = 15;

    outfnameprefix2 = filename;
```

2) **Read input files**

The input files specified on the command line are read in this phase. The software automatically computes the number of banks based on the dimension of the data in the first input file and allocates the corresponding data structures accordingly.

```
/* count number of banks (nrows in data matrix) */
nbanks = 0;
while (fgets(buffer, MAXSTRLEN, fp)) {
    nbanks++;
}
rewind(fp);

gsl_matrix * m = gsl_matrix_alloc (1, nbanks);
gsl_matrix * c = gsl_matrix_alloc (nbanks, nbanks);
gsl_matrix * out = gsl_matrix_alloc (1, nbanks);

/* allocate data matrix */
if ((data = (double **)malloc(nbanks*sizeof
(double *))) == NULL) {
    fprintf(stderr, "could not allocate data\n");
    exit(-1);
}
for (i=0; i<nbanks; i++) {
    if ((data[i] = (double *)malloc(NCOLS*sizeof
(double))) == NULL) {
    fprintf(stderr, "could not allocate data[%d]\n", i);
    exit(-1);
    }
    memset(data[i], 0, NCOLS*sizeof(double));
}
```

3) **Initialize data structures**

Data structures allocated in the previous step need to be initialized. Namely, the "data" matrix is initialized based on the data read from Inputfile:

```
/* reading input file and initializing data matrix */
    for (i=0; i<nbanks; i++) {
```

```
    if (fgets(buffer, MAXSTRLEN, fp) == NULL) {
        fprintf(stderr, "could not read line %d
in input file\n", i);
        exit(-1);
    }
    if (sscanf(buffer, "%lf %lf %lf %lf %lf %lf %lf
\n", &data[i][0], &data[i][1], &data[i][2], &data[i]
[3], &data[i][4], &data[i][5], &data[i][6]) < NCOLS) {
    fprintf(stderr, "could not read the %d tokens expected
\n", NCOLS);
    exit(-1);
    }
  }
  fclose(fp);
```

Here the "data" matrix has nbanks rows and NCOLS (fixed to 7) columns and holds the values coming from the "input file."

4) **Allocate data structures**

Once the number of banks (nbanks) involved in the simulation is known, all relevant data structures that are required in the computation and to hold output values can be allocated and initialized to zero:

```
    /* allocating matrix BankLoss */
    if ((BankLoss = (double **)malloc
(nTotDefaults*sizeof(double *))) == NULL) {
        fprintf(stderr, "could not allocate BankLoss\n");
        exit(-1);
    }
    for (i=0; i<nTotDefaults; i++) {
        if ((BankLoss[i] = (double *)malloc
(nbanks*sizeof(double))) == NULL) {
          fprintf(stderr, "could not allocate BankLoss
[%d]\n", i);
          exit(-1);
        }
        memset(BankLoss[i], 0, nbanks*sizeof(double));
    }

    /* allocating RContrib */
```

```
   if ((RContrib = (double *)malloc(nbanks*sizeof
(double))) == NULL) {
      fprintf(stderr, "could not allocate RContrib\n");
      exit(-1);
   }
   memset(RContrib, 0, nbanks*sizeof(double));

  /* allocating EcLoss */
   if ((EcLoss = (double *)malloc(nTotDefaults*sizeof
(double))) == NULL) {
      fprintf(stderr, "could not allocate EcLoss\n");
      exit(-1);
   }
   memset(EcLoss, 0, nTotDefaults*sizeof(double));

  /* allocating DIScharged */
   if ((DIScharged = (double *)malloc
(nTotDefaults*sizeof(double))) == NULL) {
      fprintf(stderr, "could not allocate DIScharged\n");
      exit(-1);
   }
   memset(DIScharged, 0, nTotDefaults*sizeof(double));

  /* allocating ECcharged */
   if ((ECcharged = (double *)malloc
(nTotDefaults*sizeof(double))) == NULL) {
      fprintf(stderr, "could not allocate ECcharged\n");
      exit(-1);
   }
   memset(ECcharged, 0, nTotDefaults*sizeof(double));

  /* allocating market */
   if ((market = (double *)malloc(nbanks*sizeof
(double))) == NULL) {
      fprintf(stderr, "could not allocate market\n");
      exit(-1);
   }
   memset(market, 0, nbanks*sizeof(double));

  /* allocating corr */
   if ((corr = (double *)malloc(nbanks*sizeof
(double))) == NULL) {
```

```
        fprintf(stderr, "could not allocate corr\n");
        exit(-1);
    }
    memset(corr, 0, nbanks*sizeof(double));

    /* allocating BankDefaulted */
    if ((BankDefaulted = (int *)malloc(nbanks*sizeof
(int))) == NULL) {
        fprintf(stderr, "could not allocate BankDefaulted\n");
        exit(-1);
    }
    memset(BankDefaulted, 0, nbanks*sizeof(int));

    /* allocating BankDefaultedOld */
    if ((BankDefaultedOld = (int *)malloc(nbanks*sizeof
(int))) == NULL) {
        fprintf(stderr, "could not allocate BankDefaulted
Old\n");
        exit(-1);
    }
    memset(BankDefaultedOld, 0, nbanks*sizeof(int));
```

The meaning of the above data structures in the simulation code is as follows:

BankLoss
Each row (iteration) holds the total loss for each bank. This matrix is nTotDefaults X nbanks in dimension.

RContrib
Records the risk contribution for each bank above the preset "large loss limit" threshold. It is an array of dimension nbanks.

EcLoss
Records the value of excess losses result of banks defaulted in the simulation. It is an array of dimension nbanks.

DISCharged
Records the value of deposits to be covered as a result of banks defaulted in the simulation. It is an array of dimension nbanks.

ECcharged

Records the excess losses of banks defaulted in the simulation. It is an array of dimension nbanks.

BankDefaulted

Records which bank defaults in the simulation. It is an array of dimension nbanks.

5) **Compute correlation matrix**

During this step, the "corr" correlation matrix is initialized and afterward we compute its Cholesky decomposition, using the linear algebra GPL-licensed software GSL.

```
/* compute correlation matrix and its Cholesky decomp */
       for (i = 0; i < nbanks; i++) {
              corr[i]=pow((rho+data[i][6]), 0.5);
              }
       for (i = 0; i < nbanks; i++) {
              for (j = 0; j < nbanks; j++) {
              gsl_matrix_set (c, i, j, corr[i]*corr[j]);
              if (i==j) {gsl_matrix_set (c, i, j, 1);}
              }
       }
gsl_linalg_cholesky_decomp(c);

    for (i = 0; i < nbanks; i++) {
           for (j = 0; j < nbanks; j++) {
                  if (i<j) {gsl_matrix_set (c, i, j, 0);}
                  }
           }
```

6) **Monte Carlo loop**

The main computation phase starts with an outer Monte Carlo simulation, which is carried out until a certain number of scenarios have been computed (100.000 in the default value). After each iteration, if at least one default has occurred, the variable nDefaults is increased, and the iteration values are recorded. The Monte Carlo simulation therefore proceeds until nDefaults equal nTotDefaults (100.000). This means that the total number of iterations is not known a priori, as it depends on a bank's stability: the higher the stability, the higher the number of iterations due for completing the simulations set. When no default has occurred, no values are recorded.

```
/* loop until number of defaults equals nTotDefaults */
nDefaults = 0;
niter = -1;
while (nDefaults < nTotDefaults) {
   niter++;

   hadDefault = 0;
   memset(BankDefaulted, 0, nbanks*sizeof(int));
```

BankDefaulted is an array of dimension nbanks that will record a 1 for the banks defaulted in the current iteration, and therefore is initialized to zero.

6.1 Generate random bank losses
A random number is generated for each bank:

```
for (b=0; b<nbanks; b++) {
gsl_matrix_set (m, 0, b, gsl_ran_gaussian(r,1));
}
```

Random values are then correlated among them:

```
for (j = 0; j < nbanks; j++){
      tmp=0;
      for (i = 0; i < nbanks; i++){
               tmp+=gsl_matrix_get(m, 0, i)
*gsl_matrix_get(c, j, i);
               }
      gsl_matrix_set (out, 0, j, tmp);
      }
```

The random correlated value is transformed into simulated unitary loss by means of the FIRB formula:

```
for (i=0; i<nbanks; i++) {
      market[i]= gsl_matrix_get (out, 0, i);
         }
```

```
for (b=0; b<nbanks; b++) {
  Baseloss = regcap(data[b][0], market[b], LGD);
```

And the random correlated value is then multiplied by the bank dimension (total assets):

```
BankLoss[nDefaults][b] = data[b][3]*Baseloss;
```

And the simulated loss is compared to capital for verifying if the bank has defaulted:

```
/* Check for defaults  */

if (BankLoss[nDefaults][b] > data[b][2]) {
          BankDefaulted[b] = 1;
          hadDefault = 1;
      }
    }
      for (j=0; j<nbanks; j++) {
          BankDefaultedOld[j]=BankDefaulted[j];
          }
```

6.2 Interbank simulation

After checking each of the banks independent of each other, it is of paramount importance to properly account for interbank lending and possible contagion. Unless the step is skipped by the user by means of the command-line option, the simulation is performed here. An inner loop propagates the loss to all other banks and checks if some other bank defaults as a result of contagion. The loss is then propagated to the system and the whole iteration continues until the system stabilizes (there are no more contagion defaults) or all banks default.

```
if (interbanksimulation) {
  cont = "c";
  allDefaulted = 1;
  for (j=0; j<nbanks; j++) {
     if (BankDefaultedOld[j] == 0) {
        allDefaulted = 0;
```

```
            break;
        }
    }


    hadDefaultInterbank = hadDefault;
    while ((allDefaulted==0) && (hadDefaultInterbank)) {
        hadDefaultInterbank = 0;
        InterBankDefault = 0;
        InterBankLoss = 0;
        for (j=0; j<nbanks; j++) {
            if (BankDefaulted[j]) {InterBankDefault += data
[j][4];}
InterBankLoss += data[j][5];
        }

        InterbankUnitaryLoss = (InterBankDefault/
InterBankLoss);
        if ((InterBankDefault/InterBankLoss)>1)
{InterbankUnitaryLoss = 1;}

for (j=0; j<nbanks; j++) {
        BankDefaulted[j] = 0;
        BankLoss[nDefaults][j] += (InterbankUnitary
Loss*data[j][5]);
        if ((BankLoss[nDefaults][j] > data[j][2]) &&
(BankDefaultedOld[j] == 0)) {
                    BankDefaulted[j] = 1;
                    hadDefaultInterbank = 1;
                    BankDefaultedOld[j] = 1;
                    }
            }
        allDefaulted = 1;
        for (j=0; j<nbanks; j++) {
            if (BankDefaultedOld[j] == 0) {
                    allDefaulted = 0;
                    break;
            }
            }
        }
    }
    /* end of interbank simulation */
```

At the end of the interbank contagion simulation, all data structures are updated for the next iteration of the outer Monte Carlo
loop:

```
/* update data structures before next iteration */

for (j=0; j<nbanks; j++) {
    if (BankDefaultedOld[j]) {
            numDefaultedBanks++;
            DIScharged[nDefaults] += data[j][1];
            ECcharged[nDefaults] += (BankLoss
[nDefaults][j] - data[j][2]);
    }
}

if (hadDefault) {
    nDefaults++;
}

/* printf("Iterations=%d, nDefaults=%d/%d,
nIBDefaults=%d\n", niter+1, nDefaults,
numDefaultedBanks); */
    }

intsize4 = floor(log10(numDefaultedBanks+1)) + 1;
```

7) **Dump all output files**

   Once the Monte Carlo simulation is over (the number of
   defaults equals nTotDefaults), the loop ends and all output
   values are dumped to disk.
The output values that are generated are as follows:

```
dump BankLoss:    nbanks
dump EContrib:    nbanks      > LargeLossLimit
dump EcLoss       nDefaults nbanks    > LargeLossLimit
dump DIScharged   nDefaults

    /* dump BankLoss */
    if (dumpbankloss) {
        outfname1 = malloc(strlen(outfnameprefix1)
  + strlen("_") + strlen(cont) + strlen("_") + strlen
```

```
(outfnameprefix2) + intsize1 + strlen("_") + intsize2 +
strlen("-")+ intsize4 + strlen(".dat") + 1);
     sprintf(outfname1, "%s_%s_%s_%d-%d-%d.dat",
outfnameprefix1, outfnameprefix2, cont, LargeLossLimit,
numDefaultedBanks, niter+1);
     if ((ofp1 = fopen(outfname1, "w")) == NULL) {
         fprintf(stderr, "could not open %s for
writing\n", outfname1);
          exit(-1);
     }
     for (i=0; i<nTotDefaults; i++) {
         fprintf(ofp1, "%1.3lf", BankLoss[i][0]);
         for (b=1; b<nbanks; b++) {
             fprintf(ofp1, " %1.3lf", BankLoss[i][b]);
         }
         fprintf(ofp1, "\n");
     }
     fclose(ofp1);
  }

  /* dump EContrib */
  if (dumpecontrib) {
     outfname1 = malloc(strlen(outfnameprefix3)
 + strlen("_") + strlen(outfnameprefix2) + strlen("_")
 + strlen(cont) + strlen("_") + intsize1 + strlen
("_") + intsize2 + strlen("-") + intsize4 + strlen
(".dat") + 1);
     sprintf(outfname1, "%s_%s_%s_%d-%d-%d.dat",
  outfnameprefix3, outfnameprefix2, cont, LargeLossLimit,
  numDefaultedBanks, niter+1);
     if ((ofp3 = fopen(outfname1, "w")) == NULL) {
  fprintf(stderr, "could not open %s for writing\n",
  outfname1);
  exit(-1);
     }
     for (i=0; i<nDefaults; i++) {
             if (ECcharged[i] > LargeLossLimit) {
     for (b=0; b<nbanks; b++) {
           if (BankLoss[i][b]>data[b][2]) {
     RContrib[b] += ((((ECcharged[i] -
  LargeLossLimit)/ ECcharged[i])*(BankLoss[i][b]-data
  [b][2])) / (niter+1));
```

```
                            }

                        }

                    }
            }


    for (i=0; i<nbanks; i++) {
      fprintf(ofp3, "%f\n", RContrib[i]);
    }
    fclose(ofp3);
    }


    /* dump EcLoss */
    if (dumpecloss) {
        outfname1 = malloc(strlen(outfnameprefix4)
 + strlen("_") + strlen(outfnameprefix2) + strlen("_")
 + strlen(cont) + strlen("_") + intsize1 + strlen
("_") + intsize2 + strlen("-")+ intsize4 + strlen(".
dat") + 1);
        sprintf(outfname1, "%s_%s_%s_%d-%d-%d.dat",
 outfnameprefix4, outfnameprefix2, cont, LargeLossLimit,
 numDefaultedBanks, niter+1);
        if ((ofp4 = fopen(outfname1, "w")) == NULL) {
    fprintf(stderr, "could not open %s for writing
\n", outfname1);
        exit(-1);
        }
        for (i=0; i<nDefaults; i++) {
                if (ECcharged[i] > LargeLossLimit) {
                        for (b=0; b<nbanks; b++) {
                            BankLossCap=((BankLoss[i][b] -
 data[b][2]) < (data[b][1] + data[b][4]) ? (BankLoss
[i][b] - data[b][2]):(data[b][1] + data[b][4]));
 if (BankLoss[i][b]>data[b][2]){EcLoss[i]+=
 BankLossCap;}
                        }

                    }
                }

    for (i=0; i<nDefaults; i++) {
```

```
        fprintf(ofp4, "%f\n", EcLoss[i]);
    }

    fclose(ofp4);
    }


    /* dump DIScharged */
    outfname2 = malloc(strlen(outfnameprefix2) + strlen
("_") + strlen(cont) + strlen("_")
 + intsize1 + strlen("_") + intsize2 + strlen("-")
 + intsize4 + strlen(".dat") + 1);
    sprintf(outfname2, "%s_%s_%d-%d-%d.dat",
outfnameprefix2, cont, LargeLossLimit,
numDefaultedBanks, niter+1);
    if ((ofp2 = fopen(outfname2, "w")) == NULL) {
        fprintf(stderr, "could not open %s for writing
\n", outfname2);
        exit(-1);
    }

    for (i=0; i<nTotDefaults; i++) {
        fprintf(ofp2, "%f\n", DIScharged[i]);
    }
    fclose(ofp2);

    return 0;
}
```

References

Acerbi, C., Nordio, C., and Sirtori, C. (2001) Expected shortfall as a tool for financial risk management. *Journal of Emerging Market Finance*, **8** (3), 87–107.

Acerbi, C. and Tasche, D. (2002) On the coherence of expected shortfall. *Journal of Banking & Finance*, **26**, 1487–1503.

Acharya, V.V., Drechsler, I., and Schnabl, P. (2011) A phyrrhic Victory? Bank Bailouts and Sovereign Credit Risk. CEPR Discussion Paper 8679.

Acharya, V., Engle, R., and Richardson, M. (2012) Capital shortfall: a new approach to ranking and regulating systemic risks. *American Economic Review Papers and Proceedings*, **102**, 59–64.

Acharya, V., Pedersen, L.H., Philippon, T., and Richardson, M. (2010) Measuring Systemic Risk. Technical Report, Department of Finance, NYU.

Adrian, T. and Brunnermeier, M.K. (2011) CoVaR. Federal Reserve Bank of New York Staff Reports No. 348.

Albertazzi, U. and Gambacorta, L. (2009) Bank profitability and the business cycle. *Journal of Financial Stability*, **5** (4), 393–409.

Alexander, V. and Anker, P. (1997) Fiscal discipline and the question of convergence of national interest rates in the European Union. *Open Economies Review*, **8**, 335–352.

Allen, F., Babus, A., and Carletti, E. (2012) Asset commonality, debt maturity and systemic risk. *Journal of Financial Economics*, **104** (3), 519–534.

Allen, F. and Gale, D. (2000) Financial contagion. *Journal of Political Economy*, **108**, 1–33.

Altman, E.I. (1977) Predicting performance in the savings and loan association industry. *Journal of Monetary Economics*, **3**, 443–466.

Altman, E.I. and Saunders, A. (1998) Credit risk measurement: developments over the last 20 years. *Journal of Banking & Finance*, **21**, 1721–1742.

Altman, E.I., Brady, B., Resti, A., and Sironi, A. (2005) The link between default and recovery rates: theory, empirical evidence, and implications. *Journal of Business*, **78**, 2203–2227.

Altman, E.I., Cizel, J., and Rijken, H.A. (2014) *Anatomy of bank distress: the information content of accounting fundamentals within and across countries.* Available at http://ssrn.com/abstract=2504926.

Altman, E., Resti, A., and Sironi, A. (2005) *Recovery Risk*, Risk Books, London.

Altman, E.I. and Vellore, M.K. (1996) Almost everything you wanted to know about recoveries on defaulted bonds. *Financial Analysts Journal*, 57–64.

Ang, A. and Longstaff, F. (2013) Systemic sovereign credit risk: lessons from the U.S. and Europe. *Journal of Monetary Economics*, **60**, 493–510.

Angelini, P., Mariesca, G., and Russo, D. (1996) Systemic risk in the netting system. *Journal of Banking and Finance*, **20**, 853–868.

Athanasoglou, P.P., Brissimis, S.N., and Delis, M.D. (2008) Bank-specific, industry-specific and macroeconomic determinants of bank profitability. *Journal of International Financial Markets, Institutions and Money*, **18** (2), 121–136.

Bank of England (2010) Financial Stability Report. Issue No. 27.

Barrios, S., Iversen, P., Lewandowska, M., and Setze, R. (2009) *Determinants of intra-euro area government bond spreads during the financial crisis.* European Economy, Economic papers 388.

Basel Committee on Banking Supervision (2005) An Explanatory Note on the Basel II IRB Risk Weight Functions.

Basel Committee on Banking Supervision (2006) International Convergence of Capital Measurement and Capital Standards.

Basel Committee on Banking Supervision (2010a) A Global Regulatory Framework for More Resilient Banks and Banking Systems (revised 2011).

Basel Committee on Banking Supervision (2010b) The Transmission Channels Between the Financial and Real Sectors: A Critical Survey of the Literature. Issue No. 27.

Basel Committee on Banking Supervision (2013a) Global Systemically Important Banks.

Basel Committee on Banking Supervision (2013b) Revised Basel III Leverage Ratio Framework and Disclosure Requirements.

Basel Committee on Banking Supervision (2013c) Global Systemically Important Banks: Updated Assessment Methodology and the Higher Loss Absorbency Requirement.

Beale, N., Rand, D.G., Battey, H., Croxson, K., May, R.M., and Novak, M.A. (2011) Individual versus systemic risk and the regulator's dilemma. *Proceedings of the National Academy of Sciences of the United States of America* **108**, 31.

Bernoth, K. and Erdogan, B. (2010) Sovereign bond yield spreads: a time-varying coefficient approach. *Journal of International Money and Finance*, **31** (3), 639–656.

Bernoth, K. von Hagen, J. and Schuknecht, L. (2004) Sovereign Risk Premia in the European Government Bond Market. European Central Bank Working Paper Series No. 369.

Bikker, J.A. and Hu, H. (2002) Cyclical Patterns in Profits, Provisioning and Lending of Banks. DNB Staff Reports, No. 86, Amsterdam.

Blien, U. and Graef, F. (1997) Entropy optimisation methods for the estimation of tables, in *Classification, Data Analysis, and Data Highways* (eds I. Balderjahn, R. Mathar, and M. Schader), Springer, Berlin, pp. 3–15.

Borgy, V., Laubach, T., Mésonnier, J.S., and Renne, J.P. (2011) in *Fiscal Sustainability, Default Risk and Euro Area Sovereign Bond Spreads*, Mimeo.

Breuer, T., Jandacka, M., Rheinberger, K., and Summer, M. (2010) Does adding up of economic capital for market and credit risk amount to conservative risk assessment? *Journal of Banking & Finance*, **34** (2010), 703–712.

Brownlees, C.T. and Engle, R. (2012) Volatility, Correlation and Tails for Systemic Risk Measurement. New York University Working Paper.

Bruche, M. and Gonzalez-Aguado, C. (2010) Recovery rates, default probabilities, and the credit cycle. *Journal of Banking & Finance*, **34** (4), 754–764.

Bruche, M. and Suarez, J. (2010) Deposit insurance and money market freezes. *Journal of Monetary Economics*, **57** (1), 45–61.

Brunnermeier, M.K., Crocket, A., Goodhart, C., Perssaud, A., and Shin, H. (2009) The Fundamental Principles of Financial Regulation. 11th Geneva Report on the World Economy.

Brusco, S. and Castiglionesi, F. (2007) Liquidity coinsurance, moral hazard and financial contagion. *Journal of Finance*, **62**, 2275–2302.

Campolongo, F., De Lisa, R., Zedda, S., Vallascas, F., and Marchesi, M. (2010) *Deposit Insurance Schemes: Target Fund and Risk-Based Contributions in Line with Basel II Regulation*, EUR: Scientific and Technical Research Series, European Commission Publication Office, Luxembourg.

Cannas, G., De Lisa, R., Galliani, C., and Zedda, S. (2012) *The Role of Contagion in Financial Crises: An Uncertainty Test on Interbank Patterns*. EUR: Scientific and Technical Research Series, European Commission, Luxembourg, Publication Office of the European Union.

Castiglionesi, F. (2007) Financial contagion and the role of the central bank. *Journal of Banking & Finance*, **31**, 81–101.

Castro, C. and Ferrari, S. (2014) Measuring and testing for the systemically important financial institutions. *Journal of Empirical Finance*, **25**, 1–14.

Cecchetti, S.G., Kohler, M., and Upper, C. (2009) Financial Crises and Economic Activity. Working Paper, BIS.

Codogno, L., Favero, C.A. and Missale, A. (2003) Yield spreads on EMU government bonds. *Economic Policy*, **37**, 503–532.

Cole, R.A. and White, L.J. (2012) Déjà Vu all over again: the causes of U.S. commercial bank failures this time around. *Journal of Financial Services Research*, **42**, 5–29.

Committee on the Global Financial System (2011) The Impact of Sovereign Credit Risk on Bank Funding Conditions. BIS CGFS Paper No. 43.

Co-Pierre, G. (2013) The effect of the interbank network structure on contagion and common shocks. *Journal of Banking & Finance*, **37**, 2216–2228.

Copeland, L. and Jones, S.A. (2001) Default probabilities of European sovereign debt: market-based estimates. *Applied Economics Letters*, **8** (5), 321–324.

Craig, B. and von Peter, G. (2014) Interbank tiering and money center banks. *Journal of Financial Intermediation*, **23**, 322–347.

De Lisa, R., Zedda, S., Vallascas, F., Campolongo, F., and Marchesi, M. (2011) Modelling deposit insurance schemes' losses in a Basel 2 framework. *Journal of Financial Services Research*, **40** (3), 123–141.

Degryse, H. and Nguyen, G. (2007) Interbank exposures: an empirical examination of contagion risk in the Belgian banking system. *International Journal of Central Banking*, **3**, 123–171.

Demirguc-Kunt, (1989) Deposit-institution failures: a review of empirical literature. *Economic Review*, **25**, 2–11.

Demirguc-Kunt, A. and Huizinga, H. (2000) Financial Structure and Bank Profitability. The World Bank, Policy Research Working Paper Series 2430.

Dietrich, A. and Wanzenried, G. (2011) Determinants of bank profitability before and during the crisis: evidence from Switzerland. *Journal of International Financial Markets, Institutions and Money*, **21** (3), 307–327.

Drehmann, M., Sorensen, S., and Stringa, M. (2010) The integrated impact of credit and interest rate risk on banks: a dynamic framework and stress testing application. *Journal of Banking & Finance*, **34** (2010), 713–729.

Drehmann, M. and Tarashev, N. (2013) Measuring the systemic importance of interconnected banks. *Journal of Financial Intermediation*, **22**, 586–607.

Duffie, D. and Singleton, K.J. (1999) Modeling term structures of defaultable bond yields. *Review of Financial Studies*, **12**, 687–720.

Duffie, D. and Singleton, K.J. (2003) *Credit Risk: Pricing, Measurement, and Management,* Princeton University Press.

Edwards, S. (1986) The Pricing of Bonds and Bank Loans in International Markets: An Empirical Analysis of Developing Countries' Foreign Borrowing. NBER Working Paper No. 1689.

Elsinger, H., Lehar, A., and Summer, M. (2006a) Using market information for banking system risk assessment. *International Journal of Central Banking,* **2,** 137–165.

Elsinger, H., Lehar, A., and Summer, M. (2006b). Risk assessment for banking systems. *Management Science,* **52,** 9.

Eschenbach, F. and Schuknecht, L. (2002) The Fiscal Costs of Financial Instability Revisited. European Central Bank Working Paper No. 191.

Estrella, A. and Schich, S. (2011) Sovereign and banking sector debt: interconnections through guarantees. *OECD Journal: Financial Market Trends,* **2011** (2), 21–45.

European Banking Authority (2014) Results of 2014 EU-Wide Stress Test. Released on October 26, 2014.

European Central Bank (2010) Euro Area Fiscal Policies and the Crisis. ECB Occasional Paper Series No. 109.

European Commission (2011a). *Public Finances in EMU 2011: European Economy 3,* Economic and Financial Affairs.

European Commission (2011b). Impact Assessment Accompanying the Proposal for a Directive of the European Parliament and of the Council Establishing a Framework for the Recovery and Resolution. Commission Staff Working Document, Directorate-General for Internal Market and Services.

European Commission (2014) Economic Review of the Financial Regulation Agenda. Commission Staff Working Document, Directorate-General for Internal Market and Services.

Eurostat (2013) Statistics in Focus No. 10/2013.

Favero, C. and Missale, A. (2011) *Sovereign Spreads in the Euro Area. Which Prospects for a Eurobond?* Economic Policy, Fifty-Fourth Panel Meeting Hosted by the National Bank of Poland, Warsaw, pp. 27–28.

Feng, D., Cheng, F., and Xu, W. (2013) Efficient leave-one-out strategy for supervised feature selection. *Tsinghua Science and Technology,* **18** (6), 629–635.

Fight, A. (2004) *Understanding International Bank Risk*, Wiley Finance.

Financial Stability Board (2010) *Reducing the Moral Hazard Posed by Systemically Important Financial Institutions*:FSB Recommendations and Time Lines. Financial Stability Board.

Freixas, X., Parigi, B., and Rochet, J.C. (2000) Systemic Risk, Interbank Relations and Liquidity Provision by the Central Bank. *Journal of Money, Credit and Banking*, **32**, 611–638.

Furceri, D. and Zdzienicka, A. (2010) The Consequences of Banking Crises for Public Debt. OECD Economics Department Working Papers No. 801.

Furfine, C.H. (2003) Interbank exposures: quantifying the risk of contagion. *Journal of Money, Credit and Banking*, **35** (1), 111–128.

Gabrieli, S. (2011) The Microstructure of the Money Market Before and After the Financial Crisis: A Network Perspective. Research Paper 181, CEIS.

Galliani, C. and Zedda, S. (2015) Will the bail-in break the vicious circle between banks and their sovereign? *Computational Economics*, **45**, 597–614.

Garratt, R.J., Mahadeva, L., and Svirydzenka, K. (2011) Mapping Systemic Risk in the International Banking Network. Working Paper 413, Bank of England.

Garratt, R.J., Mahadeva, L., and Svirydzenka, K. (2014) The great entanglement: the contagious capacity of the international banking network just before the 2008 crisis. *Journal of Banking & Finance*, **49**, 367–385.

Gauthier, C., Lehar, A., and Souissi, M. (2012) Macroprudential capital requirements and systemic risk. *Journal of Financial Intermediation*, **21**, 594–618.

Goddard, J., Molyneux, P., and Wilson, J.O.S. (2004) Dynamics of growth and profitability in banking. *Journal of Money, Credit and Banking*, **36** (6), 1069–1090.

Hałaj, G. and Kok, C. (2013) Assessing Interbank Contagion Using Simulated Networks. ECB Working Paper Series No. 1506.

Hasan, I., Siddique, A., and Xian, S. (2009) Characterizing the Interaction of Market, Credit and Interest Rate Risks: What the

Financial Markets Say Versus What the Banks Report to the Regulators. Working Paper.

Hasman, A. and Samartin, M. (2008) Information acquisition and financial contagion. *Journal of Banking & Finance*, **32**, 2136–2147.

Huang, X., Zhou, H., and Zhu, H. (2009) A framework for assessing the systemic risk of major financial institutions. *Journal of Banking & Finance*, **33**, 2036–2049.

Huang, X., Zhou, H., and Zhu, H. (2011) Systemic risk contributions. *Journal of Financial Services Research*, 1–29.

Hull, J. and White, A. (2004) Valuation of a CDO and an nth to default CDS without Monte Carlo simulation. *Journal of Derivatives*, **12**, 2.

Humphrey, D.B. (1986) Payments finality and the risk of settlement failure, in *Technology and the Regulation of Financial Markets: Securities, Futures and Banking* (eds A. Saunders, and L.J. White), Lexington Books, Lexington, MA.

International Monetary Fund (2009) *World Economic Outlook*, October.

International Monetary Fund (2010) *Global Financial Stability Report: Sovereigns, Funding and Systemic Liquidity*. World Economic and Financial Surveys, October.

James, C., (1991) The loss realised in bank failures. *The Journal of Finance*, **46**, 1223–1242.

Kanno, M. (2015) Assessing systemic risk using interbank exposures in the global banking system. *Journal of Financial Stability*, **20**, 105–130.

Karimzadeh, M., Akhtar, S.M.J., and Karimzadeh, B. (2013) Determinants of profitability of banking sector in India. *Transition Studies Review*, **20** (2), 211–219.

Kroese, D.P., Taimre, T., and Botev, Z.I. (2011) *Handbook of Monte Carlo Methods*, Wiley Series in Probability and Statistics, John Wiley & Sons, Inc. New York.

Kuo, D., Skeie, D., Vickery, J., and Youle, T. (2013) Identifying Term Interbank Loans from Fedwire Payments Data. Federal Reserve Bank of New York Staff Reports, No. 603.

Ladley, D. (2011) Contagion and Risk-Sharing on the Inter-Bank Market. Discussion Papers in Economics 11/10, Department of Economics, University of Leicester.

Laeven, L. and Valencia, F. (2010) Systemic Banking Crises: A New Database. IMF Working Paper No. 08-224.

Laeven, L. and Valencia, F. (2013) Systemic banking crises database. *IMF Economic Review*, **61**, 225–270.

Larch, M. and Turrini, A. (2009) The Cyclically-Adjusted Budget Balance in EU Fiscal Policy Making: A Love at First Sight Turned into a Mature Relationship, European Economy, Economic Papers 374.

Laubach, T. (2009) New evidence on the interest rate effects of budget deficits and debt. *Journal of the European Economic Association*, **7** (4), 858–885.

Lemmen, J. and Goodhart, C. (1999) Credit risk and european government bond markets: a panel data econometric analysis. *Eastern Economic Journal*, **25**, 1.

Longstaff, F. (2010) The subprime credit crisis and contagion in financial markets. *Journal of Financial Economics*, **97**, 436–450.

Lønning, I. (2000) Default premia on European government debt. *Weltwirtschaftliches Archive*, **136** (2), 259–283.

Marchesi, M., Petracco Giudici, M., Cariboni, J., Zedda, S., and Campolongo, F. (2012) Macroeconomic Cost-Benefit Analysis of Basel III Minimum Capital Requirements and of Introducing Deposit Guarantee Schemes and Resolution Funds, EUR Scientific and Technical Research Series 24603 EN, Publications Office of the European Union, Luxembourg, pp. 1–25.

Masselink, M. and van den Noord, P. (2009) The Global Financial Crisis and Its Effects on The Netherlands. ECFIN Country Focus No. 06–10.

Memmel, C., Sachs, A., and Stein, I. (2012) Contagion in the interbank market with stochastic loss given default. *International Journal of Central Banking*, **8** (3), 177–206.

Merton, R.C. (1974) On the pricing of corporate debt: the risk structure of interest rates. *Journal of Finance*, **29** (2), 449–470.

Mistrulli, P.E. (2010) Assessing financial contagion in the interbank market: maximum entropy versus observed interbank lending patterns. *Journal of Banking & Finance*, **25**, 1114–1127.

Mourre, G., Isbasoiu, G.M., Paternoster, D., and Salto, M. (2013) The Cyclically-Adjusted Budget Balance Used in the EU Fiscal Framework: An Update. European Economy, Economic Papers 478.

Niehans, J. and Hewson, J. (1976) The eurodollar market and monetary theory. *Journal of Money, Credit and Banking*, **8** (1), 1–27.

Paltalidis, N., Gounopoulos, D., Kizys, R., and Koutelidakis, Y. (2015) Transmission channels of systemic risk and contagion in the European financial network. *Journal of Banking & Finance*, **61**, S36–S52.

Pasiouras, F. and Kosmidou, K. (2007) Factors influencing the profitability of domestic and foreign commercial banks in the European Union. *Research in International Business and Finance*, **21** (2), 222–237.

Puzanova, N. and Dullmann, K. (2013) Systemic risk contributions: a credit portfolio approach. *Journal of Banking & Finance*, **37** (2013), 1243–1257.

Reinhart, C. and Rogoff, K. (2008a) *This Time Is Different: Eight Centuries of Financial Folly*, Princeton University Press, New Jersey.

Reinhart, C. and Rogoff, K. (2008b). Is the 2007 US Sub-Prime Financial Crisis So Different? An International Historical Comparison. NBER Working Paper No. 13761.

Reinhart, C. and Rogoff, K. (2009) The aftermath of financial crises. *American Economic Review*, **99** (2), 466–472.

Reinhart, C. and Rogoff, K. (2011) From financial crash to debt crisis. *American Economic Review*, **101-5**, 1676–1706.

Resti, A. and Sironi, A. (2007) *Risk Management and Shareholders' Value in Banking*, Wiley Finance.

Sahajwala, R. and Van den Bergh, P. (2000) Supervisory Risk Assessment and Early Warning System. Basel Committee on Banking Supervision Working Paper No. 4.

Shapley, L.S. (1953) A value for n-person games, in *Contributions to the Theory of Games*, vol. II (eds H.W. Kuhn, and A.W. Tucker), Annals of Mathematical Studies, vol. 28, Princeton University Press, pp. 307–317.

Sheldon, G. and Maurer, M. (1998) Interbank lending and systemic risk: an empirical analysis for Switzerland. *Swiss Journal of Economics and Statistics*, **134** 4 (2), 685–704.

Steinbacher, M., Steinbacher, M., and Steinbacher, M. (2016) Robustness of banking networks to idiosyncratic and systemic shocks: a network-based approach. *Journal of Economic Interaction and Coordination*, **11**, 95–117.

Tang, D.Y. and Yan, H. (2010) Market conditions, default risk and credit spreads. *Journal of Banking & Finance*, **34**, 743–753.

Tarashev, N., Borio, C., and Tsatsaronis, K. (2009) The systemic importance of financial institutions. *BIS Quarterly Review*, 75–87.

Upper, C. (2011) Simulation methods to assess the danger of contagion in interbank markets. *Journal of Financial Stability*, **7** (3), 111–125.

Upper, C. and Worms, A. (2004) Estimating bilateral exposures in the German interbank market: is there a danger of contagion? *European Economic Review*, **48**, 827–849.

Vuillemey, G. and Peltonen, T.A. (2013) Disentangling the Bond-CDS Nexus A Stress Test Model of the CDS Market. ECB Working Paper Series, No. 1599.

Wells, S. (2004) Financial Interlinkages in the United Kingdom's Interbank Market and the Risk of Contagion. Working Paper No. 230, Bank of England.

Zedda, S. (2015) Directs vs. side effects in financial contagion: what weights more? in *Advances in Artificial Economics* (eds F. Ambalard, F.J. Miguel, A. Blanchet, and B. Gaudou), Lecture Notes in Economics and Mathematical Systems, vol. 676, Springer, pp. 131–138.

Zedda, S. and Cannas, G. (2016) Assessing Banks' Systemic Risk Contribution: A Leave-One-Out Approach, Working Paper, available at http://ssrn.com/abstract=2687920.

Zedda, S., Cannas, G., Galliani, C., and De Lisa, R. (2012a). The Role of Contagion in Financial Crises: An Uncertainty Test on Interbank Patterns. EUR: Scientific and Technical Research Series, Publications Office of the European Union, Luxembourg.

Zedda, S., Cannas, G., and Galliani, C. (2014) The determinants of interbank contagion: do patterns matter? in *Mathematical and Statistical Methods for Actuarial Sciences and Finance* (eds M. Corazza, and C. Pizzi), Springer International Publishing, Switzerland.

Zedda, S., Cariboni, J., Marchesi, M., Petracco Giudici, M., and Salto, M. (2012b). The EU Sovereign Debt Crisis: Potential Effects on EU Banking Systems and Policy Options. EUR Scientific and Technical Research Series 25556 EN, Publications Office of the European Union, Luxembourg, pp. 1–26.

Zedda, S., Pagano, A., and Cannas, G. (2012c). A Simulation Approach to Distinguish Risk Contribution Roles to Systemic Crises. EUR: Scientific and Technical Research Series, Publications Office of the European Union, Luxembourg.

Zhang, Q., Vallascas, F., Keasey, K., and Cai, C.X. (2015) Are market-based measures of global systemic importance of financial institutions useful to regulators and supervisors? *Journal of Money, Credit and Banking*, **47** (7), 1493–1442.

Index

Banking Systems Simulation: Theory, Practice, and Application of Modeling Shocks, Losses, and Contagion, First Edition. Stefano Zedda.
© 2017 John Wiley & Sons, Ltd. Published 2017 by John Wiley & Sons, Ltd.